MARCOS COLÓN

THE AMAZON IN TIMES OF WAR

Dear Pat,

thank you for being Here
to Join us in the fight
for the indigenous
communities and the
forest!   10.17/24

Marco L.

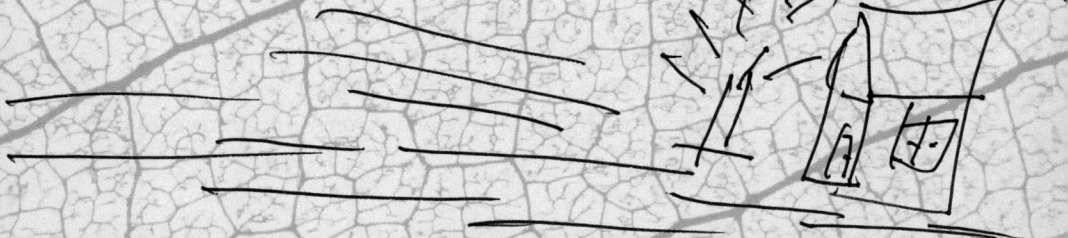

PRAISE FOR THIS BOOK

We are watching this genuine attack on a part of the world where for thousands of years people have developed ways of living inside the forest in sufficient equilibrium to produce culture and life. In the last two or three centuries it has received new inhabitants, besides the Indigenous peoples – it received colonizers, who constructed a certain humanity, learning to live within the forest. This gathering of peoples, the Indigenous and those that went there in the last two to three hundred years, makes up a large group of people living in the Amazonian forest – in the Brazilian Amazon as well as the Amazon of our neighbors Bolivia, Peru, Venezuela and Colombia. All of the Amazon basin, which affects Brazil's neighboring nations, is now shaken by this violence on the forest, but it hides another violence, which is the production of poverty, uprooting people, tearing them away from the places where they always lived supported by the forests.

The rivers and forests are constantly referred to as a place which has food, medicine, all the production of life for these peoples. They are riparians, they are extractivists of different activities, but they all depend on the cycle of production of the forests, on a cycle sensitive beyond the canopy of trees, which is the garden that exists within the forests. This garden in the forest is subsistence, it is the maintenance of life for millions of people in these countries beyond the frontiers of Brazil: all are affected and are having their lives threatened. So, I believe that beyond the concern that Europe, that the developed countries have manifested in relation to the destruction of this biome of the Amazon, the omission from this same discourse of the millions of lives suffering this violence really bothers me. As Marcos Colón demonstrates in this book, they are being displaced from their territories of origin by the fire, but also by the poverty, through a conjunction of stupidity that seems to have decided to transform prosperity into misery.

Ailton Krenak author of *Ideas to Postpone the End of the World*

Marcos Colón's powerful collection of essays traces the impact of COVID-19, Jair Bolsonaro's far-right havoc-wreaking presidency of Brazil, and the concentrated destruction of the Amazon Rainforest that accompanied these two crises. While the political scene has since shifted back to the left-wing Lula, and deforestation has begun to decline, this collection provides a poignant look-back at Bolsonaro's time in power and the multiple resistances that fought back against his policies.

Colón offers a politically engaged ecological reading of some of the most pressing issues of our time: ancestral knowledge, territory, climate, and the devastating impacts of colonialism and capitalism in one of the world's most important regions. The collection includes particularly strong essays about healthcare, the convergence of nature and culture, and the importance of taking different cultural perspectives into consideration in times of crises. The cumulative result of these essays is thoughtful insight into how public health, environmental, and Indigenous issues converge and, ultimately, the vital importance of centering the voices of Indigenous Amazonia.

Jessica Carey-Webb, Assistant professor at the University of New Mexico and the author of *Eyes on Amazonia: Transnational Perspectives on the Rubber Boom Frontier*, Vanderbilt University Press, April 2024.

Blighted by violence since the arrival of the colonizers, the Amazon is crying out for help. Marcos Colón's book is an urgent call for the action and courage needed to free the forest from plunder and pillage. This is the greatest challenge of our generation.

Cristina Serra, Brazilian journalist who worked in various media outlets, including Jornal do Brasil, Rede Globo, and Folha de São Paulo. Author of *Tragédia em Mariana: A história do maior desastre ambiental do Brasil* (Record, 2018).

We could call this book dispatches from the tropical wars against nature and its most ardent defenders — part of an ongoing imperial history — but one in a crescendo of destruction now. These essays situate the Bolsonaro period as one of acute disaster but hardly divergent from the general thrust of Amazonian transformation – a kind of alchemy of changing life into dead commodities. In this year where the Negro river has almost dried up, and the waters of the Amazon became so warm that fish were poached in their aquatic habitat, we are presented with a real set of alternatives with planetary implications. The contours of what is at stake is a vibrant socioecology on which a future could be built, or sleepwalkers' death march into the abyss.

Susanna Hetch, Professor of Urban Planning; Director, Center for Brazilian Studies at the University of California, Los Angeles. Author of *The Scramble for the Amazon and the 'Lost Paradise' of Euclides da Cunha*, The University of Chicago Press, 2013.

The Amazon has finally become political news. This outstanding collection shows why. Its defenders are being murdered, its peoples are devasted by disease and hunger, and its beautiful environment continues to be destroyed and polluted. Covering the last five or so years of this history, Marcos Colón considers the record of the Bolsonaro government and its devastating impact on the Amazon and its peoples. Yet Bolsonaro's war has failed to quash the collectivist spirit of the people who live in and from the Amazon. With stunning photographs and the sensitive ear of careful listener, Colón brings alive their voices and ambitions for a peaceful and sustainable Amazon.

> **Mark Harris**, Professor and Head of School, Humanities, University of Adelaide and Honorary Professorial Research Fellow, University of St Andrews. Author of Rebellion on the *Amazon: The Cabanagem, Race, and Popular Culture in the North of Brazil, 1798–1840*, Cambridge University Press, 2010.

This collection of topical essays, written mainly during the pandemic, should interest anyone concerned about the fate of the Amazon and its peoples. Written in an accessible and engaging style they draw on the author's encounters with people in several Amazonian countries and are interspersed with insights from the history, literature and ecology of the region.
Colón critiques the politics and corporate greed behind the destruction of the forest and its peoples, including uncontacted Indigenous tribes, and looks at how parts of the region are now dangerously out of control and in the hands of criminal gangs and drug traffickers. The section on the pandemic provides a fresh and disturbing insight into the devastating impacts of Covid-19 on many Amazonian communities due the criminal neglect of the region by the Bolsonaro government.
These essays are fresh and heartfelt and together are a powerful call for humanity to heed the voices of Indigenous peoples in the Amazon, to learn from them, and to act now.

> **Fiona Watson**, Campaigns Director, Survival International, UK.

The Amazon is in a state of agony. In 2023, it is experiencing its worst drought in history, in a way that could only be imagined by authors of the most nihilistic fiction. Images of hundreds of dead pink dolphins in dry lakes, hundreds of boats stranded on the sand, the beds of mighty rivers reduced to tiny, timid streams, a riverside village gone and buried because the absence of riverbed due to lack of rain caused the collapse of an entire cliff. All this is happening in September 2023. It is not just the effect of El Niño, but the result of political action by a former president of the Republic of Brazil. Colón's book details the origins of this tragedy and offers perspectives on what can still be done.

> **Luiz Bolognesi**, director of *Wars of Brazil* (2019) and *The Last Forest (2021)*

Marcos Colón's *The Amazon in Times of War* is an exacting, encyclopaedic account of the assault on the Amazon rainforest and its peoples during the Jair Bolsonaro regime, and beyond. It is a first step for anyone trying to understand the complexities of what has transpired there over the past decade. Especially gripping are the chapters on hunger and disease within riverside villages, and the current war on the Yanomami. In 2020, Colón was trapped on the river due to Covid travel shutdowns, and he used his journey out for first-hand observations that are unique and new. I would recommend the book to anyone concerned about the fate of the Earth's 'lungs'.

> **Joe Jackson**, Author of *The Thief at the End of the World: Rubber, Power, and the Seeds of Empire*, Penguin, 2009 and *Black Elk: The Life of an American Visionary, Farrar*, Straus and Giroux, 2016.

Marcos Colón has established himself as one of the most creative film makers to explore the politics and ecology of the Amazon. With this fine collection of wide-ranging essays, he now makes his mark as a writer too. This book is vital reading for anyone engaged by the hauntings and hopes, the fears and promise that suffuse Earth's greatest tropical forest.

> **Rob Nixon**, Professor in the Humanities and the Environment at Princeton University and affiliated with the Princeton Environmental Institute's initiative in the environmental humanities. Author of *Slow Violence and the Environmentalism of the Poor*, Harvard University Press, 2011.

Humanity and planet Earth are currently facing the greatest crisis of all time, confronting the dilemma of the survival of life. The Amazon is not merely a threatened region; it stands as the most critical and promising territory for forging other possible worlds. This is not only because of its essential role in the balance of the biosphere, but also its position as the richest space on the planet for experimenting with new forms of cohabitation of cultural diversity and biological diversity in the creative evolution of life.

Marcos Colón, through his compelling images and testimonies from the Amazonian peoples, has exposed to society the tragedy of the Amazon's destruction at the hands of the most nefarious interests of logging companies and *garimpeiros* associated with multinational corporations and the blind pursuits of global capitalism. Here, he denounces the events brought about by the direct intervention of the Jair Bolsonaro government during the pandemic times. These were clearly intended to exploit nature and decimate Amazonian communities in acts of authoritarian sovereignty leading to ecocide and the ethnocide of Amazonian territories, as well as other biomes and ecosystems in Brazil.

With refined sensitivity and critical judgment, Marcos Colón undertakes through these accounts his unwavering commitment to the defence and care of this emblematic territory home to life's diversity on the planet. An exemplary filmmaker and documentarian, he has captured in these reports the most critical processes that threaten to devastate the Amazon, while also highlighting the struggles of the Amazonian peoples who fight to both 're-exist' and coexist in their environment, to preserve and rebound their territories of life, serving as an example for all of humanity.

Enrique Leff, National Autonomous University of Mexico

Marcos Colón has done us a great service by presenting the tragic impact of the Bolsonaro presidency on the Amazon, from soaring rates of deforestation and violence against Indigenous people to the evisceration of the agencies charged with protecting the environment. Thoughtful and thoroughly reported, *The Amazon in Times of War* is a must-read for all who care about the future of the rainforest — and planet Earth.

Scott Wallace, author of *The Unconquered: In Search of the Amazons Last Uncontacted Tribes*

I have always defined capitalism in the Amazon as a war on its peoples, since generating wealth here has historically been subject to a systematic and brutal act against those that continue to produce life and communities based on common goods, as this book vividly illustrates. This war has had many adjectives through history. It was called the *guerra justa*, or just war, against the Indigenous not converted to the catholic faith in the colonial period. It was called the *correrias*, or the raid, also against the Indigenous that put themselves in the paths of the rubber tappers at the height of the *elástica* gum economy in the Amazon.

It was called development by corporate-military governments and by so-called democratic governments, which managed to dub the construction of massive hydroelectric power-stations and the neocolonial looting of minerals as progress. But this never-ending war, sharpened by this generalization of death as a strategy to make a profit, allows us to say that we are faced with a Capitalist War Against Life. In this war, some are targets, others have blood on their hands, and some, far from the battlefield, continue flaunting their delusion, imagining themselves outside the war.

Bruno Malheiro
Author of *Amazonian Horizons to Rethink Brazil and the World. Horizontes Amazônicos: para repensar o Brasil e o mundo.*

MARCOS COLÓN

THE AMAZON IN TIMES OF WAR

**Practical
ACTION
PUBLISHING**

LAB
LATIN AMERICA BUREAU

Published by Practical Action Publishing Ltd
and Latin America Bureau

Practical Action Publishing Ltd
25 Albert Street, Rugby,
Warwickshire, CV21 2SD, UK
www.practicalactionpublishing.com

Latin America Bureau (Research & Action) Ltd
Enfield House, Castle Street, Clun,
Shropshire, SY7 8JU, UK
www.lab.org.uk

A catalogue record for this book is available from the British Library.
A catalogue record for this book has been requested from the Library of Congress.

ISBN 978-1-78853-437-6 Paperback
ISBN 978-1-78853-439-0 Electronic book

Colón, M. (2024) *The Amazon in Times of War*, Rugby, UK: Practical Action Publishing and Latin America Bureau <http://doi.org/10.3362/9781788534390>.

Since 1974, Practical Action Publishing has published and disseminated books and information in support of international development work throughout the world. Practical Action Publishing is a trading name of Practical Action Publishing Ltd (Company Reg. No. 1159018), the wholly owned publishing company of Practical Action. Practical Action Publishing trades only in support of its parent charity objectives and any profits are covenanted back to Practical Action (Charity Reg. No. 247257, Group VAT Registration No. 880 9924 76). Latin America Bureau (Research and Action) Limited is a UK registered charity (no. 1113039). Since 1977 LAB has been publishing books, news, analysis and information about Latin America, reporting consistently from the perspective of the region's poor, oppressed or marginalized communities, and social movements. In 2015 LAB entered into a publishing partnership with Practical Action Publishing.

Cover and text design by: Fabricio Vinhas
Cover photo by: Edmar Barros.
Map by Patricia Tebet
Typeset by: Katarzyna Markowska, Practical Action Publishing

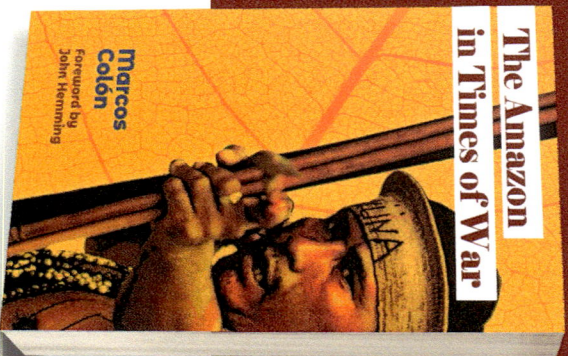

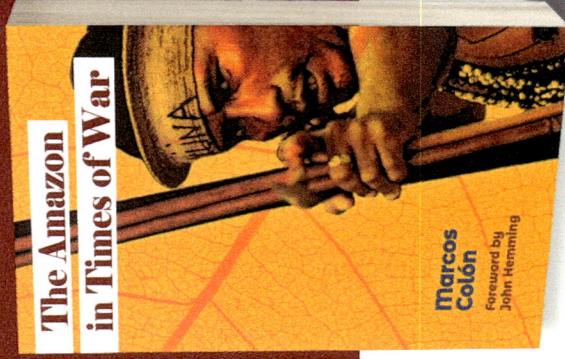

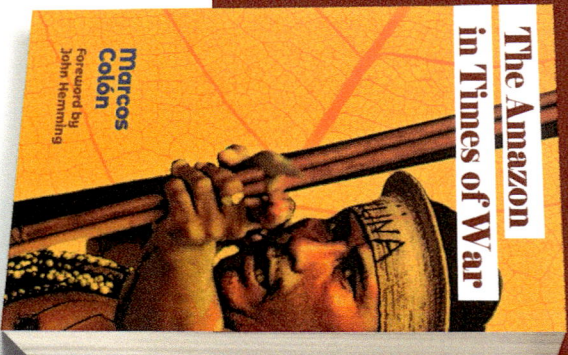

THE AMAZON IN TIMES OF WAR

By Marcos Colón

Foreword by John Hemming

'Colón has exposed the tragedy of the Amazon's destruction at the hands of the most nefarious interests of logging companies and garimpeiros, associated with multinational corporations and the blind pursuits of global capitalism.'

Enrique Leff, National Autonomous University of Mexico

For bulk order discounts email publishinginfo@practicalaction.org.uk

Order your print copy or ebook at practicalactionpublishing.com

Publish date: October 2024
9781788534376 Paperback £24.95 | $37.95 | €31.95
9781788534390 ebook £12.99 | $19.99 | €16.99
www.practicalactionpublishing.com

The Amazon in Times of War exposes the deliberate state policies behind the violence and devastation inflicted upon the Brazilian Amazon and its inhabitants.

The collection features firsthand accounts detailing not just physical assaults, but also economic and institutional harm. Zeroing in on a pivotal period from 2018, when Jair Bolsonaro assumed the presidency of an already fragmented nation, *The Amazon in Times of War* is a must-read for all who care about the future of the rainforest — and of the Earth.

The Amazon
in Times of War

Marcos
Colón

Foreword by
John Hemming

LAB
LATIN AMERICA BUREAU

Practical
ACTION
PUBLISHING

Dedicated in honor of
Dom Phillips & Bruno Pereira

Contents

Acknowledgements   xiii

List of abbreviations   xv

Foreword by John Hemming   xviii

Introduction   xxii

Part 1 - The Amazon in Times of War   **1**

I. Environmental Fascism is Haunting the Amazon   2

II. The Fire Balance Sheet   8

III. A Brief Overview of Violence in the Amazon   14

IV. The Year of Killing   22

V. Two Men Missing in the Amazon 'Wild West'   28

VI. 'Letting the Stampede Through': Changes in Environmental Laws
During the Pandemic   34

VII. Will the Amazon Rainforest Become a Commodity?   42

Part 2 - The Amazon and the Pandemic   **49**

VIII. Hunger in the Amazon: The Invisible Companion of COVID-19   50

IX. Deregulation and Deforestation Fuel the Pandemic in the Amazon   76

X. Healthcare Means Going to the Community   82

XI. Brazil's Yanomami People: Silence, Devastation, and Fear   90

XII. Above the Marombas: The Pandemic in the Amphibious Amazon   96

XIII. The Amazon and the Enigma of 'Pure Luck'   100

Part 3 - Beyond War: Life in the Amazon   **105**

XIV. A Paradise Under Suspicion   106

XV. Only a Global Coalition Will Save the Indigenous Peoples of
the Amazon   114

XVI. Amazônia Redux: A Re-evaluation of Urgent Needs   118

XVII. Stepping Softly on the Earth   122

XVIII. COP26: Cognitive Dissonance   128

XIX. Another Brazil is Possible   132

XX. Epilogue: The Amazon Is Still At War   136

Afterword by Scott Slovic: We are all Amazonians   **153**

Endnotes   158

Index   180

Acknowledgements

First and foremost, my acknowledgement goes to the Indigenous communities across the vast expanse of *Amazônia,* who have preceded me and tirelessly championed the protection of our Mother Earth. Their enduring resistance against external pressures, striving to protect and uphold their lives, land, and culture for over 500 years, has been an invaluable lesson. To those at the forefront of these challenges, know that I stand with you. It is my hope that this book will inspire the planting of seeds that, in time, will bloom into powerful catalysts for change: Decolonize. Revolutionize. Indigenize.

I express heartfelt thanks to Fiona Watson, Sue Branford, Rebecca Wilson, Mike Gatehouse and the dedicated teams at *Latin America Bureau (LAB)* and *Practical Action Publishing* for making this book a reality. To the *Amazônia Latitude* Team and the *Público* in Portugal, your support and the platform you provided to share these stories with a wider audience have been invaluable.

A special acknowledgment to John Hemming and his family for their hospitality, his enlightening foreword, and above all, the inspiration that came with every interaction. Warm appreciation also goes to Erik Jennings and his family, who always welcomed me with open arms along the banks of the Tapajós river, where many of these stories were conceived and written.

Special thanks to Scott Slovic for generously penning the afterword. From the outset, his invaluable input has significantly enriched this book project.

I am deeply indebted to everyone — including those whose names have somehow escape this list: Jessica Carey-Webb, Mark Harris, Fiona Watson, Luiz Bolognesi, Jorge Bodanzky, Cristina Serra, Joe Jackson, Scott Wallace, Fabiano Maisonnave, Jeffrey Hoelle, Enrique Leff, Suzanna Hecht, June Carolyn Erlick, José Ribamar Bessa Freire, Carlos Nobre, David Worstell, Brian Deyo and Alberto Vargas.

A shout-out to friends in Manaus: Marcus Barros, Marilene Corrêa da Silva, Ednea Mascarenhas, and Joaquim Melo, my friend who died far too young and was the first person to receive me and share his home, Amazonian experiences and librarian knowledge, and delectable *tambaquis* in Manaus. In Belém, a special 'obrigado', João de Jesus Paes Loureiro, Paulo Vieira, Márcia Collinge and James Bogan.

I am deeply grateful to Edward Layland, Chanelle Dupuis, Frances Cetti, Gabriela Isgar, Karen Bandeira, Rafael Farias, João Pires de Deus, and Lucas Lacerda. Their invaluable contributions, from thorough readings and reviews to indispensable organizational efforts, have been the backbone of this work. Without their unwavering support, this book would not have come to fruition.

As this book reached its final stages, it immensely benefited from the meticulous work of two outstanding professionals: copy editor Kim Olson and designer Fabricio Vinhas. Both Kim and Fabricio have epitomized what I envisioned in an editor and designer, respectively.

I owe a great debt of gratitude to Ailton Krenak, Lúcio Flávio Pinto, Leopoldo Bernucci, Bruno Malheiro, and Mara Régia. Despite their bustling lives, they graciously shared their insights during unforgettable conversations, illuminating discussions with both elegance and generosity.

I also honor the memory of a dear friend and mentor, Carlos Walter Porto-Gonçalves. Although he passed away before witnessing this book's completion, his ideas, inspiration, and guidance shine through every page. The insights from our conversation after he became the first to review my film, *Stepping Softly on the Earth*, profoundly influenced it. His musings on 'the colors and flavors' of Amazônia remain close to my heart.

A heartfelt thanks to Bernie for his unwavering commitment of care, and support over the years. Every day, he gently knocked on my office door, a dogged reminder to continue writing and complete each project. His constant guidance and mentorship have been my compass, and without him, this journey would not have been possible. With all my gratitude, thank you!

Finally, this one is to Chabela — compañera — who steadfastly remained by my side, ensuring that the boat of hope for this project never capsized. She is there, even when she leaves.

In addition to the people who figure in its pages, I have written this book to you, Amazônia, you beauty.

List of abbreviations

TERM	ORIGINAL PORTUGUESE	ENGLISH TRANSLATION
Amazon Triple Border	Trapézio Amazônico	Region in the Amazon where Colombia, Peru and Brazil meet
APIB	Articulação dos Povos Indígenas do Brasil	Coalition of Indigenous Peoples of Brazil
APOINME	Articulação dos Povos e Organizações Indígenas do Nordeste, Minas Gerais e Espírito Santo	Coalition of Indigenous Peoples of the Northeast, Minas Gerais e Espírito Santo
Northern Arc	Arco Norte	'Northern Belt' – According to Brazil's National Science Foundation, it refers to a regional infrastructure project seeking to transform the Amazon into a transnational shipping corridor for soybeans and corn
ASIBAMA	Associação dos Servidores da Carreira de Especialista em Meio Ambiente	Association of Civil Servants in Environmental Management
BAPEs	Bases de Proteção Etnoambiental	Ethno-Environmental Protection Bases
CASAI	Casa de Saúde Indígena	House of Indigenous Health
CIMI	Conselho Indigenista Missionário	Indigenous Missionary Council
CNA	Conselho da Amazônia	National Council of the Amazon
COIAB	Coordenação das Organizações Indígenas da Amazônia Brasileira	Coordination of Indigenous Organizations of the Brazilian Amazon
COMPAJ	Penitenciário Anísio Jobim	Anísio Jobim Penitentiary
CPI	Comissão Parlamentar de Inquérito	Parliamentary Commission of Inquiry
CPT	Comissão Pastoral da Terra	Pastoral Land Commission
CV	Comando Vermelho	'Red Commando' most prominent organized crime syndicate in Rio de Janeiro
DEHS	Delegacia Especializada em Homicídios e Sequestros	Special Division on Homicides and Kidnappings
DETER	Detecção de Desmatamento em Tempo Real	Real-Time Deforestation Detection
FARC	Fuerzas Armadas Revolucionarias de Colombia	Revolutionary Armed Forces of Colombia
FDN	Família do Norte	'Family of the North': Brazil's third largest criminal faction
FNDF	Fundo Nacional de Desenvolvimento Florestal	National Fund for Forest Development
FUNAI	Fundo Nacional dos Povos Indígenas	National Indigenous People's Foundation

GLO	Garantia da Lei e da Ordem	Law and Order Guarantee (decree)
HRW		Human Rights Watch
IBAMA	Instituto Brasileiro do Meio Ambiente e Dos Recursos Naturais Renováveis	Brazilian Institute of the Environment and Renewable Natural Resources
IBEF	Instituto de Biodiversidade e Florestas	Institute of Biodiversity and Forests
IBGE	Instituto Brasileiro de Geografia e Estatística	Brazilian Institute of Geography and Statistics
ICJ		International Court of Justice
ICMBio	Instituto Chico Mendes de Conservação da Biodiversidade	Chico Mendes Institute for Biodiversity Conservation
ILO		International Labor Organization
INPE	Instituto Nacional de Pesquisa Espacial	National Institute for Space Research
IPEA	Instituto de Pesquisa Econômica Aplicada	Institute for Applied Economic Research
IPHAN	Instituto do Patrimônio Histórico e Artístico Nacional	National Institute for Historic and Artistic Heritage
MAPA	Ministério da Agricultura, Pecuária e Abastecimento	Ministry of Agriculture, Livestock and Food Supply
MFP	Ministério Público Federal	Federal Public Prosecutor's Office
MMA	Ministério do Meio Ambiente	Ministry of the Environment
MNTB	Missão Novas Tribos do Brazil	New Tribes Mission of Brazil
MPs	Medidas Provisórias	Executive Orders
NESAM	Núcleo de Estudos Socioambientais da Amazônia	Center for Social and Environmental Studies of the Amazon
PAOF	Plano Anual de Outorga Florestal	Annual Forestry Grant Plan
PCC	Primeiro Comando da Capital	Brazil's largest criminal gang (see entry for FDN for contrast)
PDP	Plano Diretor Participativo	Participatory Master Plan
PECs	Propostas de Emenda Constitucional	Proposed Constitutional Amendments
PROAM	Instituto Brasileiro de Proteção Ambiental	Brazilian Institute for Environmental Protection
PRODES	Projeto de Monitoramento do Desmatamento na Amazônia Legal por Satélite	Satellite Deforestation Monitoring Project for the Legal Amazon
REDE	Rede Sustentabilidade	The Sustainability Network: an environmentalist political party
SESAI	Secretaria Especial de Saúde Indígena	Special Secretariat of Indigenous Health
SFB	Serviço Florestal Brasileiro	Brazilian Forest Service
SIGEF	Sistema de Gestão Fundiária	National Land Management Service
STF	Supremo Tribunal Federal	Supreme Federal Court
UNIVAJA	União dos Povos do Vale do Javari	Union of Indigenous Peoples of the Javari Valley

About the author

Marcos Colón is the Southwest Borderlands Initiative Professor of Media and Indigenous Communities at Arizona State University's Walter Cronkite School of Journalism and Mass Communication. His research focuses on Brazilian literary and cultural studies, with a particular emphasis on the Amazon, Indigenous studies, and representations of *natureculture* in documentary film and world cinema. In addition to his academic role, he is a journalist whose articles have been featured in the *Jornal Público, Folha de São Paulo, Revista Piauí, Le Club de Mediapart, Harvard Review of Latin America, Latin America Bureau, El País* and other outlets. He has also produced and directed two feature documentary films that represent diverse perspectives on humanity's complex relations with the natural world: *Beyond Fordlândia* (2018) and *Stepping Softly on the Earth* (2022). He is the editor and founder of *Amazônia Latitude*, a digital environmental magazine.

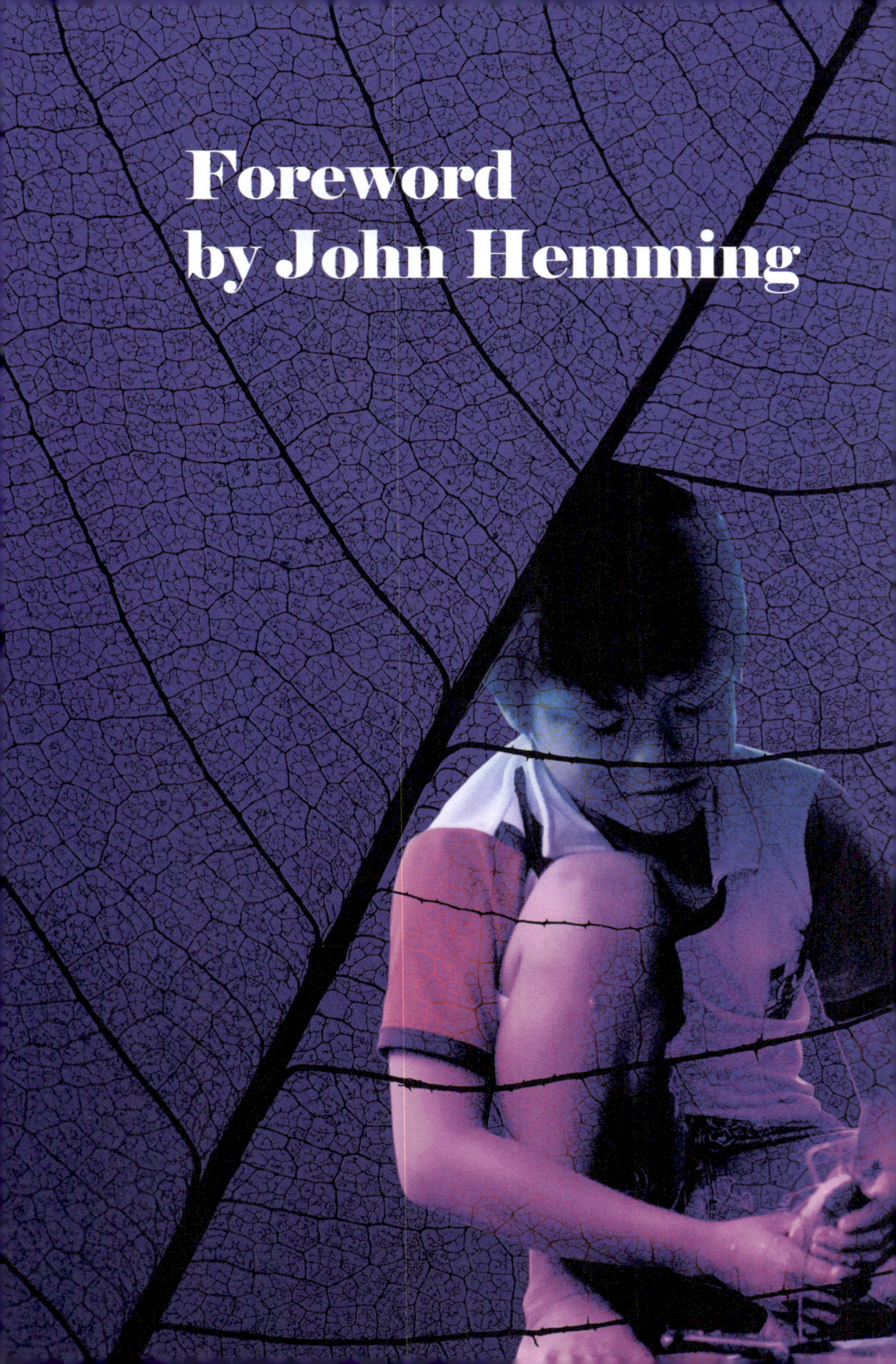

Foreword
by John Hemming

Marcos Colón is right to describe the Amazon as fighting for survival in a war of extermination. His collection of essays is timely, because these tell general readers about a mortal combat to save one of earth's most important ecosystems.

The Amazon is by far the world's greatest river, in both its sheer size and the immensity of its flow – one fifth of the freshwater that enters the oceans from all rivers. It also harbors the greatest expanse of tropical rainforest, over half the world's total, covering nine million square kilometers, the size of the continental United States.

This magnificent and beautiful environment provides three great services. Its forests generate rain for most of South America and the Caribbean, essential for agriculture and urban life. Its expanse of evergreen trees sequesters as much carbon as polluting human beings put into the atmosphere – and, conversely, if these trees were destroyed by fire or felling and rotting, they would release a tonnage of carbon that could endanger life on our planet. The third service is of less immediate benefit to mankind but is a moral imperative for all who believe in nature and creation. It is that Amazonia is the world's richest and most diverse terrestrial ecosystem, home to more plant, animal and insect species than any other biome. It is disgraceful for our one species to destroy the lives and habitats of so many millions of others with whom we share this planet.

Professor Colón is right to concentrate on recent decades because this is when Amazonia is being destroyed, almost to a tipping point from which it cannot recover.

Europeans had considered Amazonia to be a useless wilderness, during the colonial centuries after their first sighting of the river's mouth in 1500 and their first descent of it in 1542. The commodity of most interest to early colonists was the slave labor of Indigenous peoples captured from the main Amazon and its tributaries. Thus, when the Iberian kingdoms of Spain and Portugal divided most of South America between themselves at the Treaty of Madrid in 1750, most of 'worthless' Amazonia fell to Portuguese Brazil because its slavers (and missionaries) had plied the rivers in pursuit of human prey, while Spaniards on the far side of the Andes were exploiting what had been the Inca empire. By that time, most of the continent's native peoples had been destroyed by lethal imported diseases against which they had no genetic immunity. So, colonists in non-forested lands were importing millions of enslaved Africans.

The first valuable commodity from Amazonian forests was rubber from the latex of the towering hevea brasiliensis tree. The Amazon rubber boom of the late 19th and early 20th centuries brought giddy wealth to cities like Manaus, Iquitos and Belém, and misery to rubber-tapping seringueiros. But it caused minimal environmental damage because rubber trees were tapped but not felled. Amazonian rubber trees had to be tapped in the wild, because on plantations they were destroyed by South American leaf blight (SALB or *microcyclus ulei*). Thus, the Amazon rubber boom crashed abruptly when thousands of rubber trees were planted in SALB-free Malaya and what is now Indonesia. Amazonia lapsed into environmental tranquility for most of the last century.

Two inventions made it all too easy to destroy forests: chainsaws, and earth-moving bulldozers – to topple and haul trees, and to build 'penetration highways' such as the Trans-Amazonian launched by Brazil's military government in 1970. In those early days, the most sought-after Amazonian product was timber, particularly mahogany. European and North American cattle could not thrive in the intense dry and rainy seasons, with ticks and other parasites; and their lush pasture could not grow on the weak soils under destroyed rainforests. But these problems were overcome by importing humped zebu cattle and tough grasses like *colonião* from India. The next problem was to walk this livestock to slaughterhouses without losing most of the animals' meat. This was resolved by building more roads and, above all, by paving them to serve in all weathers. Sawmills and slaughterhouses also moved deep into forested Amazonia, to be close to the logs and cattle they were processing. Brazil became the world's leading exporter of beef, and much of this originated on ranches in destroyed Amazonian forests.

The third great driver of deforestation is soybeans, a plant imported from China and Japan. This is the world's only great food crop with nitrogen-fixing properties, which means it can grow on very weak soils. (Tropical rainforests are evergreen, so falling nutrients are immediately and constantly recaptured by the growing biomass. There is no winter break during which humus and topsoil can accumulate. Thus, when forests are destroyed, the exposed earth is little more than acidic sand.) Soybeans are a nutrient-dense protein. Worldwide demand for this crop soared because it is perfect animal feed as well as an additive for human foods. A road known as the Soya Highway was sliced through the heart of Brazil's forests, northwards to a deep-water soy port at Santarém on the lower Amazon[1].

Localized damage is also caused by mining, notably for iron ore, cassiterite for tin, and alluvial panning for gold. Attempts to build hydroelectric dams in the very flat expanse of Amazonia's forests have required the flooding of gigantic reservoirs, and these dams decimate freshwater fish stocks by blocking spawning migrations.

These are the financial drivers that have caused the destruction or degradation of almost half of the Amazon's seemingly endless forests. This environmental catastrophe has accelerated in recent decades, and it forms the backdrop of many chapters in this book. Another great theme is the treatment and survival of Indigenous peoples in Brazil and in the seven[2] adjacent Amazonian republics. Tribes in the more open east and south of Amazonia were decimated by disease and immigrant settlement during the colonial centuries. But forest peoples survived with their communal societies and cultures intact when they moved away from the major navigable rivers. This survival was helped by Indigenous champions, such as Marshal Cândido Rondon and then the Villas Boas brothers in Brazil, and others in surrounding countries. The result was the creation of vast areas of protected 'Indigenous Territories' throughout Amazonia, and highly favorable clauses in Brazil's post-military Constitution of 1988. Indigenous peoples were recognized as the original Brazilians. They are also seen as the best custodians of the tropical forests in which they live and which they revere.

Introduction

This collection of essays provides evidence (both direct and indirect) that the violence and devastation wreaked on the Brazilian Amazon and its peoples is due to deliberate state policy. The essays, comprising first-hand descriptions of not only physical aggression but economic and institutional violence, were originally published through various academic and media channels at a time of increasing awareness of an impending threat: a political scheme to destroy the world's largest biome, one that covers a total of nine countries in South America. Many of the worst aspects of this threat were deployed or exacerbated by the Bolsonaro government and form the major focus of this book.

I have resisted the temptation to update the essays in the light of events subsequent to the end of the Bolsonaro presidency. Minor editing, brief introductions and endnotes are used to explain the context in which they were written and to correct or clarify a few details. But I must ask the reader to accept them substantially as they were written at the time, without the benefit of hindsight. I have, however, added a brief epilogue chapter that sets out some of the policies for the Amazon adopted by the incoming administration of Luis Inácio Lula da Silva, their shortcomings and the challenges Brazil now faces if it is to reverse the devastation wrecked in the Bolsonaro years.

The essays explore various themes but follow a basic chronology, united by their focus on a particular moment beginning in 2018, that reflects many of the unresolved traumas of the country's past: the scourge of slavery, the genocide of its Indigenous peoples, and the devastation of its environment. In November of that year, after a second round of voting in the Brazil's general elections, Jair Bolsonaro took office as president of a deeply divided country. The world watched as a little-known conservative federal deputy, former army captain and neoliberal enthusiast, was elected head of state of the largest nation in South America. His rhetoric (crudely) mirrored that of his apparent North American counterpart, to the extent that the media labelled him the 'Trump of the Tropics'.

The start of the Bolsonaro administration was a unique moment in which this political plan invoked, reinforced (often with a fascist bias), and made visible the power exercised over the Amazon and its peoples throughout Brazil's history. Bolsonaro became the avatar for the exercise of explicit violence: now no longer faceless, it operates beyond what Rob Nixon[3] refers to as the 'slow violence' of environmental degradation, or the less visible violence that emerges from a diachronic process which marginalizes and abuses economically peripheral groups and racial minorities.

This plot has its own goals and, occasionally, practical effects. For example, after he was expelled from the army, Bolsonaro became a Member of Congress with the aim of fighting the demarcation of Indigenous lands and authorizing their potential availability as places for commercial exploitation, such as the extraction of minerals or timber. When he came to power, these were just two of the many measures the new president and his right-hand man, Minister of the Environment Ricardo Salles, tried to force or pass unnoticed through Congress or, failing that, to inject into an already inflamed public debate.

To understand the nature of the threat to the Amazon represented by the neoliberal economic and agro-industrial complex, one first needs to reflect on how the region is literally burning – and burning far more than usual. In large part, these flames have been fanned by the former president through an irresponsible policy that sought to create enemies and invert roles, blaming those who have historically been responsible for keeping the Amazon Forest standing. Bolsonaro claimed that native and traditional peoples of the Amazon were responsible for the destruction of their own habitat.

Indeed, according to the Bolsonaro administration, the blame does not lie with its own specific actions, despite all evidence to the contrary. For example, there is the crisis of the R$187 m (about £30 m) within the Ministry of the Environment, which resulted in a series of cuts to Brazil's environmental protection programs, culminating in a 95 per cent cut to the budget underpinning the national policy on climate change. There were also successive attempts to sabotage Brazil's environmental surveillance agencies. Under the command of Salles, these agencies were demoralized and lost the authority to do what they were tasked to do by law: monitor and punish criminals who harm the environment, especially in the Amazon. And finally, there is the National Indigenous Foundation (FUNAI), the agency responsible for mapping and protecting Indigenous lands, whose administration was radically undermined during Bolsonaro's time in power.

It is worth remembering, however, that none of this is really new. For years, centuries even, megalomanic projects aimed at 'taming' and developing what has been called by Alberto Rangel and Euclides da Cunha the 'Green Hell' have impacted life in the Amazon. Nor has this been exclusive to Brazil. Nevertheless, dams, mining, deforestation, aggressive economic and infrastructure projects (to cite only the legal ventures) have grown into an increasingly large and out-of-control amalgam of threats.

Of course, it is important to note that such irresponsible projects find their backing even further afield, in the economic powers outside Brazil that drain materials, resources and energy from the Amazon to keep the wheels of capitalism turning.

In the first year of the Bolsonaro administration, these acts of violence escalated. The fires in the Amazon were more intense, lasted longer, destroyed more of the forest, and had repercussions on the global environment. During this period, many more Indigenous people were expelled in the name of development, and many more were killed. This was no coincidence. Prior to this, the rate of killings was already alarming, with frequent murders of community and environmental leaders. Tensions in the field were not tempered during Bolsonaro's time in office; on the contrary, attacks carried out by environmental criminals increased, empowered by the president's coded call to arms, as well as by institutional indifference and financial incentives.

All these problems were magnified during the coronavirus pandemic, which began in 2020 and continues to have widespread social, political, and economic fallout. The disease gained a particularly destructive force in the humid lands of the Amazon, where the organizational and infrastructural problems of the region, combined with the historic indifference of the Brazilian government, led to a chaotic horror show. The world witnessed the spectacle of human lives in the Amazon needlessly sacrificed in droves at the height of the outbreak.

These essays, however, are not limited to a critique of this destructive model of development, whose sustenance relies on the devastation of both human and nonhuman lives. There are also encounters with those who have resisted and continue to resist this devastation, despite the threat of violent death at the hands of destructive and criminal forces empowered by the Bolsonaro administration. By paying attention to these voices we arrive at other ways, other routes to safeguard the Amazon, its peoples, and even the global ecosystem.

Yet one of the effects of the murders, fires, and destruction we see in the Amazon is the silencing of those voices that cry out against this process. To try to counter this, we need to begin a dialogue that sheds light on what has historically remained in the shadows. Talking to and learning from the Amazonian peoples, particularly in such difficult and desperate times, teaches us that there are other ways of doing things, ways that these peoples have been practising for thousands of years.

This book not only tells stories of destruction and violence, and its emergence in the Amazon as a state-run neoliberal project, aggravated by the outbreak of the coronavirus pandemic, but also celebrates the voices of its subjects, looking to them for positive reinforcement, to keep alive the hope of other possibilities for the Amazon. These are the voices that, as Ailton Krenak[4] writes, postpone the end of the world. They know much better than we do the way out of the crisis of civilization in which we find ourselves. It is also for them that I have collected these essays into book form.

SOUTH AMERICA

BELÉM

TOCANTINS RIVER

BELO MONTE

PARÁ

XINGU

SANTARÉM

VALE

ITAITUBA

TAPAJÓS RIVER

MANAUS

RORAIMA

YANOMAMI TERRITORY

NEGRO RIVER

MADEIRA RIVER

RONDÔNIA

AMAZONAS

SOLIMÕES RIVER

ACRE

LETICIA

TABATINGA

JAVARI VALLEY

IQUITOS

NANAI RIVER

Part 1

The Amazon in Times of War

I. Environmental Fascism is Haunting the Amazon [5]

This essay was published in 2018 – a symbolic year for one of the first and last utopias on Planet Earth: the Amazon rainforest. In that year, a spectre prowled the region, in the person of Jair Messias Bolsonaro, candidate for the presidency of Brazil. He went on to win the second round of the elections on 28 October and became President of Brazil on 1 January 2019.

An area of over 400 hectares of forest burned in Amazonas, August 2021. Photo: Edmar Barros/ Amazônia Latitude

Presidential candidate Jair Bolsonaro's ignorance of the Amazon lays bare the poverty of the economic and political ideas he champions. Besides the rainforest's importance in terms of Brazil's socio-cultural and historical diversity, its integration with hydrography presents a complex set of problems and solutions not only for the country but for the entire world. Yet for Bolsonaro, the Amazon is just an undeveloped, remote area of Brazil, 'a place of [ignorant] Indians, Quilombolas[6] and Caboclos[7]' that needs to give way to progress. Bolsonaro's idea of 'progress' or 'development' is represented by chopping down the forest and clearing the way for the advance of agribusiness, funded by foreign investors. In announcing to the world that the 'Amazon is not ours' and 'we have ways to explore this region in partnership', Bolsonaro ignores the evidence of its Indigenous peoples' adaptability over millennia, of all the latest scientific thought on the region, of the resistance and resilience of the humid tropical cultures, and of the fact that the Amazon forest is shared with eight other countries besides Brazil – Venezuela, Colombia, Peru, Bolivia, Ecuador, Suriname, Guyana, and French Guyana.

Open for business

At a business meeting held after announcing his candidacy, Bolsonaro indicated that he favored exploration for, and exploitation of, the riches of the Amazon, whose natural resources, including minerals and oil, are valued at up to US$5 trillion. Indeed, executives participating in the event confirmed Bolsonaro's receptiveness to 'putting the Amazon back on the agenda' – the agenda, that is, of those with predatory interests. However, economic exploitation of the Amazon, as proposed by the candidate, faces a great deal of resistance from certain segments of society. In 2017, then-President Michel Temer tried to open up Renca (an area in the Amazon originally demarcated by the military government for exploitation of mineral resources), but was forced to drop his plans due to the widespread negative response.

Bolsonaro's discourse is similar to that of Donald Trump: he has announced his opposition to the global climate agreement known as the Paris Agreement. Approved by 195 countries in 2015, the Agreement aimed to reduce greenhouse gas emissions to avoid further global warming. Yet Bolsonaro believes Brazil would have to 'pay a high price' to meet the Agreement's demands, claiming it undermines the country's sovereignty:

> *What is in play is national sovereignty because there are 136 million hectares that we've lost control of. ... I will leave the Paris Agreement if this continues being an object of it. If our part means handing over 136 million hectares of the Amazon, then I'm out.*[8]

Once elected, Bolsonaro has promised to begin discussions in Brazil's Congress on the demarcation of Indigenous lands and potential commercial exploitation of protected areas. This would have implications for the sustainable use of land by riparian and Indigenous agricultural communities in the Amazon, threatening the lifestyles of native populations. Recently, Bolsonaro stated:

> *There won't be any more demarcations of Indigenous land. We're going to give a rifle and a carry permit to every farmer.*[9]

Previously, as a federal lawmaker in 1998, he expressed admiration for the ruthless methods used by the American cavalry against Native Americans during the United States' expansion.

> *The Brazilian cavalry was very incompetent. The North American cavalry were the competent ones because they decimated their Indigenous people in the past and today, they don't have this problem in their country.*[10]

Bolsonaro has been a persistent critic of the land demarcation process established by the 1988 Constitution, which set aside large areas as Indigenous reserves. In August 2019 he described the Indigenous territories as outdated, saying, 'Why must we keep them in reserves, as though they were animals?' At the same gathering he disparaged the designation of Indigenous territory as an 'industry' that needs to be stopped, adding, 'Indigenous people don't lobby, don't speak our language, and yet today they manage to have 14 percent of our national territory... One of their intentions is to hold us back.'[11]

These statements reflect a particular aspect of Bolsonaro's policies and discourses – directly threatening the Amazonian lands and its peoples – that could be referred to as environmental fascism. Historically speaking, fascist regimes are, first and foremost, considered anti-democratic: the relationship between state and society leaves no room for social mediation, and any debate or divergence of opinion from state orthodoxy is not allowed. Protests, the expression of opposing views and active participation by civil society are suppressed, and free press and media are eliminated, while power and choice become concentrated in the hands of the few.

The environmental fascism in Bolsonaro's discourse emerges in his defense of the predatory use of natural resources, which he would place under exclusive government control, and his declarations that outside opinions regarding the fate of these resources emanating from community councils, academia or NGOs, for example, are inadmissible[12]. He plans to reinforce his discourse through legal interventions aimed at extinguishing any activist forces intent on defending the Amazon, particularly rural social movements.

Then there is Bolsonaro's declared intention to restrict land use by minorities or traditional groups, and impose economic zoning on these unique and fragile ecosystems. This will disqualify precisely those whose traditional practices help limit predatory development and open up their lands to extractive activity directed by outsiders, which will increase negative environmental impacts and chances of an environmental catastrophe. This presages the minimization or even extinction of Brazil's entire post-dictatorship regulatory framework which was intended to limit the exploitation of natural environments.

There is also a risk of undermining advanced scientific research that highlights the fundamental importance of the Amazon's rich natural, cultural, and historical heritage to all humanity. This knowledge challenges the validity of claims of national sovereignty over these territories, which are often seen as unacceptable interference by a centralized, all-powerful state. Under Bolsonaro, this heritage would viewed exclusively through the lens of a narrow economic perspective supposed to serve 'national interests'. These national interests, imposed from above, operate with a fascistic logic, excluding all other interpretations and visions of the world.

Disregarding the Constitution

Statements made by Bolsonaro show little consistency or diplomatic content and demonstrate that the candidate already disregards Brazil's sovereign right and duty to participate in preservation of the region as a whole. This goes against the Brazilian Constitution of 1988, whose Article 255, paragraph 4, establishes the fundamental regulatory framework for the country:

> *Everyone has the right to an ecologically balanced environment, which is a public good for the people's use and is essential for a healthy life. The government and the community have a duty to defend and to preserve the environment for present and future generations.*

Bolsonaro's disdain and lack of knowledge concerning Indigenous populations represents a profound danger to Brazilians in the Amazon Basin. This is because it revives an old but ever-present threat. In the past, the military dictatorship and earlier conservative governments repeatedly considered opening up the Amazon to allow big business to conduct predatory explorations in order to pay off Brazil's foreign debt. Each time, however, the Amazonian populations, alongside the scientific community and well-informed Brazilian institutions, denounced this false discourse of 'national salvation,' and fought to safeguard the integrity of the land and its resources for present and future generations.

The Amazon, with all its physical, environmental, sociocultural, and historical components, is a crucial part of today's complex world. It is an intricate web of ecosystems, nestled within a biome that is essential to maintaining the Earth's equilibrium. The region is considered critical for sustaining the chemical balance of the atmosphere and the dynamics of the hydrological cycle, and for limiting climate change. Its ecosystems have the highest level of biodiversity anywhere on the planet. As such, its urban, rural, and Indigenous areas along with environmentally protected reserves should only be explored and developed in a sustainable manner, based on the constitutional order and legal framework agreed upon by the Amazon's states.

Bolsonaro represents a clear and present danger to any such sustainable development, as shown when he announced on national television:

> *We want an end to this industry of fines levied by the Brazilian Institute for the Environment and Renewable Natural Resources [IBAMA] and the Chico Mendes Institute for Biodiversity Conservation [ICMBio]. We are going to put a complete stop to all activism in Brazil*[13]

Bolsonaro does not speak for the landless, the leaseholders, or the small farmers, but has clearly made a commitment to the country's agricultural capitalists. Not only does this demonstrate his ignorance, but it underscores his disrespect for the true value of the Amazon and its importance in the lives of its Indigenous peoples and of ordinary Brazilians. He fails to understand that millions of Brazilians see the Amazon as a place where the relationship between nature and culture is emblematic of the country; its conquest through the simple reproduction of colonialism stands in direct opposition to the political struggle of its peoples. Brazilians from north to south, traditional populations, the Indigenous, quilombolas, pantaneiros[14], riparians[15], caboclos, and others have already reached the consensus that protection of the Amazon is as important as the defense of public policies such as *Fome Zero* (Zero Hunger) and *Bolsa Família* (the Family Allowance).

In this context, fake news and social networks operate as a supplementary political force, reinforcing obscurantism and the fundamentalism of racial ideologies, prejudices of every sort, and ignorance about the sustainability of environmental protection policies in the Amazon. By stating he will disband such state institutions as IBAMA and ICMBio, set up as environmental protection agencies, and do away with all 'political activism', Bolsonaro is effectively ripping up the Brazilian Constitution – which any presidential candidate is duty-bound to protect.

The spectre of Bolsonaro and his environmental fascism is deeply concerning. The Brazilian Trump shows a complete disregard for the cultural history of the very people he purports to represent, and not only is it a tragedy for the Brazilian people of the Amazon, but it is also contrary to the environmental interests of the entire world.

II. The Fire Balance Sheet [16]

This essay, written in late 2019, expresses the sense of foreboding felt by many who witnessed an environmental catastrophe unfold in record time.

A suspect environmental policy

Over the past year, Brazil's Minister of the Environment, Ricardo Salles, has been embroiled in a series of environmental controversies. Brazilians have watched in bewilderment as respected environmental agencies such as the Brazilian Institute of the Environment and Renewable Natural Resources (IBAMA), the Chico Mendes Institute for Biodiversity Conservation (ICMBio), and the National Indigenous Foundation (FUNAI) saw their staff and duties drastically reduced, losing their authority and with it their voice, silenced by the developmental discourse of the Bolsonaro administration. We heard Salles discredit the memory of Chico Mendes[17], witnessed his refusal to accept the satellite data of the National Institute for Space Research (INPE) which reveals the extent of deforestation in the Amazon, watched as he compromised the Amazon Fund[18] by his statements, and looked on in disbelief as he posed for photos wearing a traditional headdress alongside a group of Indigenous people.

Meanwhile, in parallel, although gaining far less attention, the Bolsonaro administration drew up an agreement with the US allowing economic exploitation of the Amazon. The lack of media hype was surprising, given this agreement was Bolsonaro's flagship policy, central to his economic agenda. While public attention was focused on pension reform, international economic partnerships – such as joining the *Organization for Economic Cooperation and Development* (OECD), and the agreement between the Southern Common Market (Mercosur) and the European Union – and even leaks by *The Intercept*[19] concerning ethical violations by Brazilian courts, the environment seemed to take a back seat, despite being the area most affected by government policy in the first half of 2019.

It took an international scandal and the threat of reprisals by European leaders before the federal government and the media gave the Amazon the attention it deserved. Satellite images and photos of fires raging in the Amazon Forest circulated around the world, sparking a new diplomatic crisis for Bolsonaro. The threat by the French President Emmanuel Macron to suspend the agreement between Mercosur and the European Union, in addition to international threats of boycotts and an embargo of Brazilian agricultural commodities, were what finally forced the government to tone down its most egregious policies.

Forest fires and an international crisis

At approximately 3 p.m. on 19 August 2019, the city of São Paulo was plunged into complete darkness – day became night. The first theory to gain traction was that it was due to the convergence of smoke generated by out-of-control fires burning in Central-West Brazil and the Amazon, and although it has since been contested, this theory continues to be accepted among European media as a valid explanation for the phenomenon. Indeed, a preliminary study by the University of São Paulo indicated burned biomass (material of plant origin) had been found in rainwater collected that same day. But even if there was no direct relationship with the fires, the apocalyptic atmosphere drew attention to the environmental crisis.

Over the next few days, the event would come to dominate international news cycles, elevating Brazil to the role of environmental villain. A number of world leaders accused the country of inept management of the Amazon, commenting that Brazil was on a suicidal path. Images of the raging fires in the Amazon raced around the world, lighting up newspapers and social networks, and sparking the hashtag #PrayForAmazonia.

In due course, the French president, faced with a worrisome crisis of popularity at home, took advantage of the moment to try to appear presidential on the world stage by taking the lead, proposing a veto of the Mercosur agreement and embargos on Brazilian products – this coincidentally fitted the demands of the French agriculture sector which fears the agreement's effect on its profits. However, despite listing the proposal as a priority on the G7 summit agenda, Macron found little support among the other leaders. British Prime Minister Boris Johnson, in particular, had no intention of closing the doors to Brazilian trade, since he needed to guarantee an economic solution to Britain's exit from the European Union and the loss of European markets. Nevertheless, during the summit, the establishment of a $20 million contingency fund was announced to tackle the fires in the Amazon and help alleviate the damage.

However, the exchange of barbs between Brazilian and French presidents created a hostile environment that did not bode well for implementation of the measure – Bolsonaro would only agree to accept the aid if Macron apologized for the personal offense he had caused, particularly in relation to the issue of Brazilian sovereignty over the Amazon.

Defending Amazonia

Controversies and misunderstandings aside, while heads of state were occupied with preserving their pride, the Amazon continued to burn. Up to the time of this article's publication, few concrete measures had been implemented to resolve the problem, but among those few was the Law and Order Guarantee decree (GLO), which mandated sending Brazil's National Guard to fight the fires. The international community, in turn, was ready to donate resources to help contain the crisis.

However, inhabitants of the Amazon region most immediately affected by the lack of environmental control were not consulted. For example, Porto Velho, the capital of Brazil's Rondônia State, suffered from its proximity to the fires. According to a report in *National Geographic*[20] published 23 August 2019, the city's airport had to be closed against the spread of the fire, palm trees at the entrance to the city were singed by the flames, and smoke invaded local shopping centers. The article also stated that '[the] number of people admitted to state hospitals with pneumonia, severe coughs, and other respiratory problems has tripled in the last week, according to local reports.'[21]

The Tenharin Indigenous people, inhabitants of the middle course of the Madeira River, in the southern portion of the state of Amazonas, have always looked after their land, warding off invasion, and until 24 August had seen no fires in their territory. In an interview with the *O Estado de São Paulo* newspaper[22], Indigenous leader Antônio Enésio Tenharin declared: 'We take care of this land, our territory. Until today, there has never been a fire. But now it has come to several places all at once. It is dreadful for our people, because it makes our children sick, kills the animals, and brings only bad things.' The report recorded his indignation at the actions of the government, who appeared to support the incursions of loggers, land grabbers, and prospectors, increasing the pressures on his people.

Meanwhile, Mura Indigenous leader Raimundo Mura,[23] told the G1 online portal that he would 'fight to his last drop of blood' in defense of his land. The Mura people, who today boast around 15,000 members, have a long history of resistance. They have secured the demarcation of their lands, situated at the confluence of the Madeira, Purus, and Amazon Rivers, where they follow their ancestral traditions, living off fishing and sustainable extraction. Now, however, the Mura find themselves threatened by the advancing destruction in the Amazon. Raimundo stated:

> *What is being done here is an atrocity against us. Nobody expected this to happen, but it is happening, and we are really feeling [the effects]. For years, we have resisted here, when there was no access by road, when electricity arrived, [and] when the invasion took place.*[24]

Nevertheless, despite the fires, Bolsonaro again declared during a meeting with the state governors of what is known as the Legal Amazon[25] on 27 August that he intended to continue to regulate prospecting on Indigenous lands. At the Palácio do Planalto (the presidential residence) where the meeting took place, he called the demarcation of Indigenous lands by previous governments 'irresponsible,' since demarcation makes it impossible to take economic advantage of these states. *Folha de São Paulo* observed that at the meeting, which was broadcast live on the president's social networks, Bolsonaro spoke more to the camera than to the governors, indicating that he intends to govern as if he were still on the campaign trail.

The toxicity of Ricardo Salles

Since the crisis began, the Minister of the Environment has seen even his allies distance themselves. As the main target of politicians, the press, and environmentalists, Salles has proved toxic to anyone who has dared stay close to him. Indeed, it seems his very position is now under threat. Since 20 August, when the Brazilian Institute for Environmental Protection (PROAM) and fifty other NGOs filed a request for an enquiry into administrative impropriety on the part of the ministry, Salles has been amassing political failures.

On 22 August, NOVO, the political party that includes Salles as a member, issued a public notice claiming the minister 'had not been recommended by NOVO and, therefore, does not represent the institution. The minister was chosen by and answers to the president, Jair Bolsonaro.' On the same day, REDE[26] (the Sustainability Network) went to the Supreme Court to demand Salles' impeachment, alleging he had committed the crime of irresponsibility by failing to fulfil his 'constitutional duty to protect the environment and the international commitments undertaken by Brazil.'

On 23 August, Salles blamed the fires on environmental mismanagement by previous administrations. He stated the fires had occurred because there are no legal means of economic development in the region, forcing farmers to resort to illegal measures, burning down the forest in order to make a living. The next day, in an interview with *O Estado de S.Paulo*, he continued in the same vein, adding that '[t]his story that [the Amazon] belongs to humanity is nonsense,' and rebutting international criticism.[27]

Again, on 24 August, he maintained this bravado on his social networks, tweeting: 'More fire in Angola and Congo than in the Amazon.. .and Micron [sic] says nothing.. .why is that? Could it be because they don't compete with the inefficient French farmers?'

Despite his insistence on remaining at his post, Salles is finding himself cornered – his political situation depends directly on resolving the environmental crisis, but this seems a distant prospect, particularly because it is subject to the administration's ideological crusade[28]. Meanwhile, in Brasilia, a Parliamentary Commission of Inquiry (CPI) to investigate the increase in the number of fires raging in the Amazon looms on the horizon. The proposed CPI, put forward by Randolfe Rodrigues (REDE-AP), already has the support of 27 senators and may be put in place at any time, according to Mônica Bergamo of *Folha de São Paulo*.[29]

III. A Brief Overview of Violence in the Amazon[30]

This article, written in 2019, uncovered some deeply disturbing trends in human rights violations in the Amazon.

The Tabatinga airstrip is so hidden within the vastness of the Amazon rainforest that, when arriving by plane to the city, it is natural to fear momentarily that the pilot has made a mistake. After an initial moment of panic, however, an urban setting opens up, whose tranquility is only disturbed by the comings and goings of motorcycles, the main mode of transportation. This apparent peace seems strange at first to those who have prepared themselves to enter a stronghold of Brazilian drug trafficking, located along the Amazon triple border of Brazil, Colombia, and Peru.

The team of journalists I was traveling with was there to cover the International Seminar on Political Ecology in May 2019. Prior to our arrival, our contact with the city had been through Brazilian media reports, such as 'Favela Amazônia', an article[31] from the *O Estado de S.Paulo* newspaper on the misery in Tabatinga. Videos produced by the paper showed Indigenous people scavenging for food in an open dump – a testament to the absence of even minimal services in Brazilian cities. Another clipping[32], from the *Sputnik* website, referred to the Amazonian municipality as 'a lawless outpost', and spoke about the influence of drug trafficking on daily life in Tabatinga. The feature was illustrated with a photograph of corrugated iron fencing spray-painted with graffiti by the Família do Norte (FDN) drug faction, claiming possession over the territory and announcing, 'The border is ours'.

The FDN first gained visibility with the wave of rebellions that hit Brazilian prisons in 2015. These made public the war between the Primeiro Comando da Capital (PCC) from São Paulo, the largest criminal faction in the region, and the Comando Vermelho (CV), the main faction active in the metropolitan region of Rio de Janeiro. In alliance with the CV, the FDN butchered and beheaded several PCC members detained in prisons in the north in order to secure control over drug trafficking routes in the region. The factional conflict culminated in the massacre at the Anísio Jobim Penitentiary Complex (COMPAJ), located in the city of Manaus, 1 January 2017. It was the second most violent massacre in the history of Brazil's prison system, accounting for 56 deaths.

Given the FDN's history of extreme violence, visiting Tabatinga was, at the very least, anxiety-provoking. To our surprise, upon landing in the city, which is accessible only by air or water, we were soon confronted with the same sort of graffiti as seen in *Sputnik*. We did not expect to find the FDN's 'handiwork' on Avenida da Amizade, the city's main thoroughfare, which connects it to the municipality of Leticia in Colombia. However, even here, in downtown Tabatinga, the inhabitants seemed unfazed by the graffiti, and life went on smoothly, contradicting media depictions of a violent, lawless municipality. The impression of calm would be soon dispelled, however, when a motorcycle taxi driver told us it was impossible to cross at the border post one evening as criminals had left a body there.

Even near this Amazon triple border, the region that accounts for the largest volume of cocaine production in the world, Tabatinga did not seem to suffer directly from the activities of drug trafficking. Luiz Fábio Paiva, in his article[33] 'At the margins of the nation state: discourses of violence on the Amazon Triple Border', appears to confirm this perception. Paiva conducted several interviews with residents of Tabatinga in an attempt to understand their attitude towards the violence, and shows that the bloody trail of trafficking-related murders (a gift to media exhibitionism) is not the only form of violence to befall Amazonia's populations. In fact, many of Paiva's interviewees claimed that only those directly involved in crime met violent deaths. For example, one civil police officer reported:

> *It works like this. You buy some drugs, so like, you buy on credit. Then, all of a sudden you go off with the drugs, either because the police arrest you or because you want to cheat the guy. Then they call and say, 'Either you pay for my drugs or you die'. And that's how it works, if you dont pay, you die. You can run away, but never come back here, because the day you come back, it may have been 10 years since it happened, you think they forgot, but they don't forget. When you set foot here, they usually already know you're coming, there's already a gunman waiting for you. They kill you then and there. That's how it is.[34]*

However, Amazonians have to deal with other, less explicit forms of violence, such as land grabs, colonization of nature through the illegal predatory extraction of natural resources, human trafficking, and the expansion of the agricultural frontier. The latter is occurring in a region where the state has so far authorized the use of 439 types of pesticides. Among other problems, this directly affects physical health, moral integrity, interior spaces, and the very ways of life of traditional communities. As in Paiva's work, several other articles refer to this violence using the term 'socio-environmental conflicts,' referring to occurrences of confrontation between the interests of local populations and other social agents, usually from outside. This confrontational atmosphere is not restricted to people; it also permeates the economic development model imposed on many other regions throughout the world, one that prioritizes profits over social welfare.

Socio-environmental conflict

Pedro Rapozo, a professor at Amazonas State University (UEA) and coordinator of the research entity known as the Center for Social and Environmental Studies of the Amazon (NESAM), states that socio-environmental conflicts are also marked by inequalities in political and economic capital. In the Brazilian case, both are represented by power disputes between the legislative and executive branches related to territorial demarcation policies and their delayed justice:

> *Socio-environmental conflicts maintain an agency relationship with available natural resources. The mediation of these resources is represented by the impacts generated by human action in these environments. Conflicts are also considered through divergent lifestyle values, non-consensual ideologies between hegemonic and non-hegemonic groups that dispute power for a particular good or predisposed natural resource.*[35]

In a July 2019 article[36], Rapozo explains that the competition for power generates social and environmental clashes between agents such as the state, landowners, expropriated territories, squatters, traditional fishermen, farmers, Quilombolas, and Indigenous peoples. He notes that the occurrence of conflicts, motivated by lifestyle changes that threaten the permanence and very survival of rural Amazonian societies, has intensified with the advance of agribusiness and mega-enterprises in the region:

We can clearly see this. On the one hand, civil society is made up of social groups, peasants, rural workers, squatters, fishermen, and Indigenous people. On the other, there are state actions, such as policies to finance large economic ventures. These are examples of how this happens in the Amazon. In fact, there is big capital on one side and civil society on the other. It does not mean that, within civil society itself, there are no conflicts of interest between fishermen, Indigenous peoples, riparian populations, and Quilombolas, and it is not one specific group that is especially impacted by these conflicts, considering they are the result of direct interest in the territory, its occupation and use for some good. But given the context of the Amazon, we clearly have very strong opposition on the one hand from rural societies being criminalized, and on the other hand, the expansion of ventures in the Amazon as a result of the influence of the state.[37]

Violence in numbers

September 2019 saw the publication of two reports on violence in the Amazon. The first[38], by Human Rights Watch (HRW), discusses how violence and impunity have accelerated the pace of deforestation in the region. Meanwhile, the compilation[39] by the Indigenous Missionary Council (CIMI) scrutinizes violence perpetrated against Indigenous peoples.

In *Mafias do Ipê: Violence and Deforestation in the Amazon*[40], HRW examines 28 murders, most dating from 2015, as well as four attempted murders, and more than 40 death threats. HRW investigations found that in most cases, the crimes were directly linked to illegal activities by loggers, prospectors, and squatters in the region. Among the victims were Indigenous people and members of other traditional communities who reported the criminals, as well as public officials engaged in monitoring and combating these activities.

This violence takes place with impunity: the state simply fails to pursue or punish the criminals involved in the murders, threats, and extortion. According to the HRW report, among the more than 300 murders analysed by the Pastoral Land Commission (CPT), a non-governmental organization that maintains a record of complaints not processed by the official bodies having jurisdiction in the case, only 14 of the perpetrators were tried, and of the 28 murders investigated by HRW, only two went to court, with none of the 40 death threats examined ever going to trial. This impunity is largely due to inadequate police investigations. Local police forces admit the shortcomings but attribute them to the fact that these deaths occur in remote areas. HRW, however, has documented serious omissions, such as the lack of autopsies in investigations of deaths in the city that often take place not far from a police station. As Rapozo states:

[It is not just] a matter of murders and attempted murders as physical violence that can be considered a consequence of the conflict. The imposition of certain practices, certain state policies, the lack of ensuring land use rights, the inefficiency in addressing conflict resolution cases, and the way the state and certain economic agents deal with this relationship with impacted agents, can also be seen as a model of symbolic violence, also expressed as a result of these conflicts.[41]

Indigenous peoples, who have long played the role of monitors and defenders of the Amazon Forest, combating deforestation with warnings and denunciations, are rendered particularly vulnerable by the weakening of environmental protection agencies and laws. The report on *Violence Against Indigenous Peoples in Brazil: 2018*, published by CIMI in September 2019[42], shows that there were 109 cases of 'land invasions, illegal exploitation of natural resources and miscellaneous property damage' in 2018 compared with 96 reported in 2017. Between January and September 2019, CIMI counted 160 cases of invasion of Indigenous lands, spread across 153 territories in 19 Brazilian states.

Up to the release of the report, CIMI had observed 941 cases of violence against the property of Indigenous peoples. These are attacks related to omissions and delays in titling land, conflicts over territorial rights, illegal exploitation of natural resources, and various incidents of damage to Indigenous property. In this context, the agency paid special attention to the case of the Munduruku Indigenous Territory, where it is estimated that about 500 small-scale mines have already been installed.

In 2018, CIMI reported 110 cases of violence against individuals relating to abuse of power, death threats, various other threats, manslaughter, wilful bodily harm, racism and ethnic discrimination, attempted murder, and sexual violence. In the same year there were also 135 cases of murder of Indigenous people, 25 more than that recorded in 2017. Besides violence against Indigenous peoples due to failure of State oversight or criminal violation of right, there were also suicides, increased levels of child mortality, and deaths due to lack of medical assistance, among other causes.

As indicated by the Special Secretariat for Indigenous Health (SESAI), only partial data is available, and new cases could always come to light. Thus, it is clear that the actual situation regarding the murder of Indigenous people is even more serious than the statistics indicate.

The relationship with deforestation

In 2016, by ratifying the Paris Climate Agreement, Brazil pledged to eliminate illegal deforestation in the Amazon by 2030. From 2004-2012, deforestation had been reduced by over 80 per cent per cent, from almost 28,000 km^2 of forest destroyed per year to less than 4,600 km^2. Brazil's success in combating deforestation prior to 2012 was partly due to the use of near-real-time satellite imagery to locate and close down illegal logging sites. Another success factor was the creation of protected areas – conservation units and Indigenous lands – covering hundreds of thousands of square kilometers across the Amazon region, where legal restrictions on land and resource use help protect the forest. However, after approval of the New Forest Code in Congress, deforestation rose again in 2012 as a result of intense lobbying by the agribusiness interests. In 2018, it reached 7,500 km^2, and its extent has been even greater in 2019.

On 18 November 2019, the National Institute for Space Research (INPE) published the results of the Satellite Deforestation Monitoring Project for the Legal Amazon (PRODES), which monitors deforestation over a one-year period. The index was obtained through analysis of satellite imagery produced by PRODES, which is responsible for accounting for those areas that appear to have reached the end of the deforestation process. The numbers are 42 per cent per cent higher than the 6,833.9 km^2 predicted by Real-Time Deforestation Detection (DETER), another INPE system which monitors the Amazon region on a daily basis. Among the factors identified as responsible for the increase in deforestation in 2019 are adoption of sophisticated logging techniques, which are more difficult to identify through satellite surveillance, and budget and staff cuts at federal environmental agencies, which have reduced their ability to investigate and punish loggers.

Also highlighted in the HRW report is the government's lack of interest in investigating or even filing complaints concerning illegal forest activities. This scenario of impunity and reduced enforcement provides greater freedom for loggers, squatters, and prospectors, among other criminals, to conduct their activities and aggravate the destruction of Amazonian biomes, as well as intensifying the violence against the Indigenous people who denounce them in an attempt to protect their physical safety, way of life, and the environment. It is noteworthy that the relaxation of environmental legislation and indifference to denunciations leaves civil society powerless, with no alternative ways of intervening to prevent deforestation.

International recognition of the dangers

These reports by HRW and CIMI, along with data recently released by PRODES, show a worrying trend regarding the safety of traditional populations in the Amazon region. Throughout 2019, there have been alarming news reports every week about deforestation, land invasions, threats to and murders of Indigenous leaders and members of social movements, as well as the much-discussed fires. Among many controversial activities, we have seen the discontinuation of the demarcation process of Indigenous lands (with no expected date for their resumption), the weakening of environmental monitoring and protection agencies, the approval of more than 400 pesticides for use in agriculture, and an increase in the number of attacks by criminals on Indigenous lands and other common property rights. Many of these activities are connected to the discourse of President Bolsonaro, as reports in the mainstream media and academic journals frequently demonstrate.

In light of this scenario, the University of Oxford held an international colloquium 31 January – 2 February 2020, called 'The Amazon: Growing Violence and Worrying Trends'. This event continued the discussions held at the International Seminar on Political Ecology and was part of the activities of the AgroCultures international research network,[43] which hopes to foster scientific and cultural production, and strengthen ties between thought leaders from the Amazon and other parts of the world.

The colloquium brought Indigenous leaders and leaders of social movements into dialogue with teachers, artists, and researchers dedicated to the socio-environmental problems of the Amazon region. The main objective was to debate the legacies of the past, ask questions about responsibilities, and assess the alarming rates of violence, inequality, and food insecurity.

At last the international community was waking up to the horrific escalation of violence in the Amazon, against people, the natural environment and the agencies and laws designed to protect them.

IV. The Year of Killing[44]

By putting names to the statistics, this essay covering 2019 continues the preceding analysis of violence in the Amazon, highlighting the increase in the number of murders of Indigenous leaders.

By the end of 2019 at least 26 Indigenous leaders[45] had been killed over a period of 11 years. The Pastoral Land Commission (CPT), which only considers those deaths that occur as a result of conflicts over land, recorded seven deaths in 2019 alone – a 250 per cent increase over the previous year. The data, however, are preliminary, and the final tally for 2019 will be calculated in April 2020[46].

BAR-CHART: MURDERS OF INDIGENOUS PEOPLE AND THEIR LEADERS IN THE COUNTRYSIDE

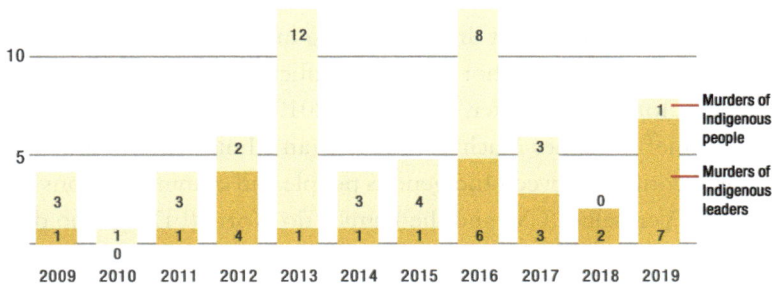

2019: Preliminary data. Source: Comissão Pastoral da Terra (Pastoral Land Commission) (CPT)

Up to 2019, the year with the highest number of recorded murders had been 2016, which saw the deaths of a total of five Indigenous leaders and eight other Indigenous people. In 2017, three leaders and three Indigenous people were killed, while in 2018, two leaders met with violent deaths. However, no sequence of years has presented such a notable difference in the number of deaths of Indigenous leaders as 2018–2019.

In total, nine Indigenous people were murdered in 2019, the last just over 40 days after the murder of a Guardian of the Forest[47], Paulo Paulino Guajajara. Erisvan Soares Guajajara (an Indigenous man) was found butchered close to the Vila Industrial District soccer field in the city of Amarante do Maranhão (683 km from the capital São Luís). The body of a non-Indigenous person was found alongside Erisvan and identified as José Roberto do Nascimento Silva. Unlike Paulo Paulino, the police stated that Erisvan was killed by drug traffickers. This was refuted by his sister Célia Lúcia Guajajara, who declared he had no connection with narco-trafficking, despite his history of substance use. The investigative journalism agency, *Amazônia Real*, verified this claim[48].

The shooting of Indigenous leaders

Seven Indigenous leaders were killed in 2019, and their deaths have generally been reported as 'due to territorial conflicts' or 'disagreements among criminal factions' , or else they have been left unresolved. The first death of the year occurred on 27 February in an attack on the Urucaia community in Manaus, when Chief Francisco de Souza Pereira (an ethnic Tukano) was shot and killed by hooded men in his own home. The motive for the crime remains unknown but the city's special police department that deals with homicides and kidnappings (DEHS) is working on two lines of investigation: the first is that this leader of 42 communities was killed for his stance against drug trafficking in the region, and the second is that he may have been shot due to a conflict over land distribution.

Almost four months later, on 13 June 2019, also in the city of Manaus (AM[49]), Chief Willames Machado Alencar (an ethnic Mura) was killed in a territorial conflict between Indigenous people and criminal factions of the Comando Vermelho (CV) and the Família do Norte (FDN), who dispute control of drug trafficking close to the Indigenous community known as Cemitério dos Índios (Indian Cemetery) in the northern zone of Manaus. Known as 'Onça Preta' ('Black Jaguar'), the Indigenous leader was shot five times during the struggle. Some say he had previously expelled his assassin from the community for involvement with drug trafficking. At the time of the crime, Chief Willames was preparing to attend a public hearing at the Amazonas Legislative Assembly in the south-central zone of the capital. According to witnesses, he had received death threats from local criminals on a constant basis.

Soon after this murder, on 22 July, Emyra Wajãpi, leader of the Wajãpi, was killed in an in an attack near Aldeia Mariry in the Wajãpi

Indigenous Territory in the municipality of Pedra Branca do Amapari, Amapá state. Community leaders stated that the killing occurred during an invasion by mining prospectors in the region. However, the Federal Public Prosecutor's Office shelved the inquiry into the murder. According to the Federal Police investigation, the death was 'accidental.' Emyra, however, was found near a river with knife wounds on his body.

On 6 August, almost two months after the killing of Chief Willames, Carlos Alberto Oliveira de Souza, known as 'Mackpak,' was killed in another attack on the Cemitério dos Índios community in Manaus. The chief, who was of Apurinã ethnicity, was shot more than 10 times. According to local residents, the motive for the crime may have been revenge, since he was one of those leading the fight against the Cemitério dos Índios land invasion and had refused to accept the presence of members of criminal factions in the area.

On 1 November, as mentioned previously, Paulo Paulino Guajajara, a member of the Guardians of the Forest, was killed in an attack on the Arariboia Indigenous land, where 92 communities of the Guajajara, Gavião, and Awá peoples are located, in Bom Jesus da Selva (MA[50]). His murder occurred during an ambush by illegal loggers. He was shot in the neck.

Just over one month later, on 7 December, two Guajajara leaders, Firmino Prexede Guajajara and Raimundo Benício Guajajara, were shot in Jenipapo dos Vieiras (MA). There is still no information concerning the perpetrators, or the motive for that crime.

Environmental policy

Months before the murder of Paulo Paulino, the Indigenous peoples of Arariboia had denounced invasions by outsiders coming to exploit the forest and the absence of any government protection. Even with the growing number of homicides among Indigenous populations, authorities have been slow to react and inefficient both in investigating the crimes and punishing the perpetrators. According to the *Amazônia Real* news agency,[51] over the last 12 years at least six Guardians of the Forest have been murdered, with no subsequent arrests or penalties. The various complaints made to state and federal authorities have met with no response. Loggers appear to have been granted impunity in their incursions into Indigenous lands and subsequent conflicts with the inhabitants.

As a protective measure, Minister of Justice and Public Security Sérgio Moro sent the National Guard to Maranhão to guarantee the safety of native peoples and employees of the National Indigenous Foundation (FUNAI), as well as non-Indigenous inhabitants. In addition, the federal government sent a representative from the Ministry for Women, the Family, and Human Rights to the Cana Brava Indigenous Territory. However, according to Paulo Tupiniquim[52], executive coordinator of the Coalition of Indigenous Peoples of Brazil (APIB), any military action orchestrated by the state brings only more insecurity, given there is no way to trust a government that spreads hatred and feeds prejudice against the native peoples of Brazil. 'It is regrettable that in the 21st century, Indigenous people are still being murdered over territorial issues, for defending their rights,' he states.

Since Jair Bolsonaro assumed power, attacks on Indigenous lands have increased, as he made it clear he would not demarcate traditional territories, has taken measures that have made environmental policy more flexible and open to abuse, and declared that he favors mineral extraction in these areas. The Indigenous Missionary Council (CIMI) has declared:

> *The recurrent statements by the President of the Republic against demarcation and regulation of territories, followed up by a regional environment prejudiced against the Indigenous, has been the main vector for invasions and violence against Indigenous peoples in Brazil.[53]*

Between January and September 2019, the CIMI report on Violence against Indigenous Peoples in Brazil described 160 cases of invasion of 153 Indigenous territories in 19 Brazilian states[54]. Bolsonaro's declarations have indeed provoked tension and an increase in violence against Indigenous peoples in Brazil. Political and economic interests have been given more importance than protection of Indigenous culture and conservation of the environment, thus promoting confusion among Indigenous populations who are trying to protect the forest, and the murder of those who challenge this situation.

Begging for help

The Indigenous peoples of the Amazon, struggling to find enough strength among their peers to pursue their struggle, are suffering a crisis due to lack of support. The government that should protect them attacks them, while a portion of the population vehemently believes in the right to exploit the Amazon and sees native peoples as obstacles to 'national progress'.

For Guenter Francisco Loebens[55], a member of CIMI's Support Team for Isolated Indigenous Peoples, it is important that Indigenous people increase their presence in various official bodies, from human rights spaces such as the United Nations and the Organization of American States (OAS) to the judicial system, the National Congress, and the Public Prosecutors' Office, in order to bolster their cause. It is also extremely important to gain the support of Brazilian society and the international community, with the aim of expanding their network of allies.

Paulo Tupiniquim, one of the coordinators of the Coalition of Indigenous People of the Northeast, Minas Gerais and Espírito Santo (APOINME), talks about the need for the government to support measures that protect Indigenous rights, such as strengthening FUNAI, demarcating and ratifying Indigenous lands, and abolishing decrees such as Executive Orders (MPs) and Proposed Constitutional Amendments (PECs) that take away the rights of Indigenous peoples and promote their direct or indirect persecution. When questioned about how the Brazilian general public could help in the Indigenous struggle, he answered that it is very important for society to believe in the cause and respect a culture that only wants its rights guaranteed:

We are fighting for our existence, fighting so that our planet remains alive. We don't want to take away anyone's rights, we don't want privileges, and we don't want to hinder the progress of the country. We just want our rights to be respected without suffering setbacks, and that Brazilian society join us in defending nature because if we continue to contaminate the land and rivers and destroy the forests, and we don't think about alternative solutions, our planet will die. Then it won't be just the Indigenous who go extinct, but all inhabitants of this planet.[56]

V. Two Men Missing in the Amazon 'Wild West' [57]

This article was originally published by Amazônia Latitude *on 28 May 2019, to describe the social problems around Tabatinga, on the border between Brazil, Colombia and Peru. It was updated on 9 June 2022, to record the disappearance and murder in the area of British journalist Dom Phillips and Indigenous expert Bruno Pereira.*

Brazil ranks third among countries with the longest international borders. It has a border that totals 16,885 km, touching every country in South America except Chile and Ecuador. The vastness of this border territory has made it difficult for public policies to address its many problems. Among the most conspicuous issues are drug trafficking, smuggling of goods, weapons and vehicles, high rates of violence and a lack of socio-economic development.

However, these problems are not found along all Brazil's borders. Every stretch has specific situations and gives rise to differing socio-environmental conflicts. The Amazon Triple Border with Colombia and Peru, in the state of Amazonas, is characterized by poor access to basic services, the presence of organized crime, human trafficking, migratory waves, criminal gangs, timber extraction and illegal mining. This is facilitated by the unusual geography of the region, which has many rivers and extensive areas of forest that make it difficult to monitor and protect national borders.

The Amazon Triple Border in the Arco Norte (Northern Arc)

The city of Tabatinga makes an interesting case study for understanding the situation on Brazil's borders. Located in the micro-region of Alto Solimões in Amazonas, Tabatinga forms a twin conurbation with the Colombian city of Letícia. Linked by the major thoroughfare known as Avenida da Amizade, the cities form one settlement, with overlapping legal jurisdictions. It requires the establishment of specific regulations for both cities, which, unfortunately, are not capable of meeting the real needs of their populations.

Data[58] from the Brazilian Institute of Geography and Statistics (IBGE) shows that only 2,700 of Tabatinga's 64,000 residents (4.8 per cent of the population) have employment documents, while 48.2 per cent survive on half the minimum wage or less per month. Only 21.6 per cent of homes in the municipality have adequate sanitation, only 8 per cent of its roads are properly paved, and the infant mortality rate is 20.88 deaths per thousand live births, much higher than the national average of 12.8.

The unusual situation on the border combined with the absence of the Brazilian, Colombian and Peruvian state organizations and Amazonian geography, facilitates the movement of drugs, weapons and other illegal products, by organized crime operating in Brazilian territory.

Drug trafficking in the south of the country is controlled by the Primeiro Comando da Capital (PCC), Brazil's largest criminal gang, but is dominated by its rival, the Família do Norte (FDN) in the north. The FDN has been directly involved in the massacres of PCC members in Brazilian prisons in recent years.

Rivalry between Brazilian drug traffickers is expressed as a struggle for domination of the illegal weapons market and the sale of drugs in domestic and international markets, as well as in their alliances with Colombian, Peruvian and Bolivian criminal organizations. Partnerships with the latter are responsible for fuelling the illegal Brazilian market, some of whose proceeds fund guerrilla groups such as the Revolutionary Armed Forces of Colombia (FARC). This scenario accounts, to a great extent, for the alarming number of violent deaths in Brazil every year: 60,000.

In addition to the 'Wild West' created by the drug traffickers, of particular note are the Indigenous people who live in the region. Socio-environmental conflicts generated by the unusual geography and state negligence have forced them to migrate to small urban settlements in search of healthcare, food and other goods. According to the daily newspaper *O Estado de São Paulo*[59] many of these immigrants are unable to return to their communities because they cannot afford the trip home. This is due to the lack of access to the city, which is limited to one small airport and rivers. There are no roads linking the municipality to Manaus, the capital of Amazonas.

A particularly worrying case is that of the Ticuna people, many of whom live in the Alto Solimões region. Their living conditions have deteriorated in the absence of state action, as described in the *O Estado* article mentioned above. According to this source, the proximity of their communities to Tabatinga has led to the development of Indigenous neighborhoods with serious social problems on the outskirts of the city. In the absence of job opportunities and basic welfare, many have to eke out a living from the city's garbage dumps and landfills, competing for food with other vulnerable sectors of the population and the vultures.

The state's lack of interest

The Institute for Applied Economic Research (IPEA) recently published its study on the Brazilian border situation. The third volume[60], entitled *Fronteiras do Brasil: uma avaliação do Arco Norte [Brazilian borders: an assessment of the Northern Arc]*, concluded that the presence of a socially and culturally diversified population in possession of sustainable land and forest management practices must be taken into account by public policies on the Amazon borders, and that development strategies adopted in other parts of the country are neither appropriate nor as effective in this environment.

However, the authors also cited a series of problems beginning with the precarious urban structure of the area and the absence of efficient transportation routes and connection to commercial centers. There is also poor communication between the various security forces and a corresponding lack of integrated strategies and information-sharing that would help maintain the security and national sovereignty of countries along the border. Problematic, too, is the Brazilian government's lack of interest in promoting economic and social development in the region, coupled with its failure to make the most of local biodiversity, preferring instead to simply administer and occupy the area. The precariousness of health in the region has also increased, as a result of poor healthcare, a lack of basic sanitation, and the presence of open landfill sites. Finally, there is the current situation of Venezuela, which has generated large waves of migrants and conflicts along the borders of several countries.

The situation along Brazil's borders demands greater attention from the government because it is directly linked to domestic and international problems such as drug trafficking.

This in turn is directly related to violence and overcrowding in prisons, ruining the lives of people who become dependent on drugs such as crack and cocaine, the latter being the main product crossing the borders in the Northern Arc. This situation highlights the need for academics, government, and social leaders, as well as civil society, to debate and solve these problems.

The Javari Valley

Located along the Amazon Brazil–Peru border is the Javari Valley Indigenous Territory, the second largest in the country, with 85,000 km^2, almost the size of Portugal. The Valley's tropical abundance has made it a point of access for illegal hunters, fishermen and loggers, provoking violent conflicts between the Indigenous inhabitants and the riverside communities[61] that opposed the establishment of the territory in 2001. It is also a smuggling route for cocaine traffickers who take advantage of the absence of government oversight and fight to control trade between Brazil, Peru and Colombia.

In the Javari Valley coca production from 2019–2020 increased by almost 20 per cent to 61,777 hectares in Peru, the second largest producer after Colombia, according to the United Nations. The growing drug trade has increased the dangers and conflicts along the triple border with Colombian and Brazilian cartels disputing control over access to the Amazon River for sending their cocaine to the lucrative European market.

The whole border region is also home to approximately 6,000 Indigenous people belonging to 26 Indigenous nations, 19 of whom live in isolation. Since assuming the presidency, Jair Bolsonaro has made such groups even more vulnerable. His government makes statements and adopts policies that facilitate extractivist activities that destroy the forest and its peoples and open up space for criminal activity. Groups that defend environmental and Indigenous rights have long argued that Bolsonaro's public position on Indigenous territories encouraged the invasion of Indigenous lands and conservation areas with impunity.

The disappearance of Dom Phillips and Bruno Pereira

It was in this area where the British journalist Dom Phillips and Indigenous rights defender Bruno Araújo Pereira disappeared. They were traveling along the Itaquaí River, the main waterway to access the Javari Valley. They were last seen when they arrived at the community of São Rafael on Sunday 5 June. They spoke there with the wife of a community leader known as 'Churrasco' before departing for Atalaia do Norte, a trip that lasts approximately two hours. They never reached their destination.

They travelled in a new boat, with a 40 horse power motor and 70 liters of gasoline, enough for the trip. The Union of Indigenous Peoples of the Javari Valley (UNIVAJA) began the search the same day, when they failed to arrive. By Monday 6 June, after failing to locate the men, UNIVAJA contacted authorities and issued a press release: 'The two men were planning to visit the Indigenous Surveillance team near the town of Lago do Jaburu (close to the FUNAI Surveillance Base on the Ituí River), where Phillips planned to interview Indigenous people.'

A suspect was arrested in connection with the disappearance, but police say there is still no evidence that a crime has been committed. *Amazônia Latitude* photojournalist Edmar Barros travelled to the region to accompany the searches.

Meanwhile, Brazil's Congress is deliberating legislation that will open up Indigenous lands for extractive industries, such as mining and timber. This will put the Javari Valley, and journalists and Indigenous rights defenders like Dom Phillips and Bruno Araújo, in even greater danger.

Postscript

As of January 2024, three people are in jail waiting trial for the murders of Bruno and Dom. However, the mastermind of the murders has not yet been formally indicted. Also the then president of FUNAI, Marcelo Xavier and another senior FUNAI official Alcir Amaral have been indicted for indirectly contributing to the murders by failing to adequately protect FUNAI workers[62].

VI. 'Letting the Stampede Through': Changes in Environmental Laws During the Pandemic [63]

In April 2020, then Minister of the Environment Ricardo Salles advised his government peers to take advantage of the media focus on COVID-19 to dismantle environmental legislation. This article, written at the time, charts what happened.

In a video released to fellow ministers on 22 April 2020, Minister of the Environment Ricardo Salles states:

> *We have the possibility, now that the attention of the media is focused almost exclusively at COVID-19, the opportunity ... to pass the infra-legal reforms of deregulation, simplification, all the reforms the world demands.*

He goes on to suggest how the government and its partners should act, 'so, for this, it needs our effort,' and adds:

> *[We should let] the stampede through and change the regulations and simplify the rules of IPHAN [the National Institute for Historic and Artistic Heritage], of the Ministry of Agriculture, of the Ministry of the Environment, of the ministry of this, of the ministry of that. Now is the time to join forces... It is regulatory change that we need, in all aspects.*

In fact, measures of deregulation, such as dismantling the Brazilian Institute of the Environment and Renewable Natural Resources (IBAMA) and the Chico Mendes Institute for Biodiversity Conservation (ICMBio), were already on course at the time. The video itself simply ratified Salles' policy of stimulating illegal logging, prospecting, land grabs, and deforestation.

On 11 April 2020, for example, following use of a Law and Order Guarantee (GLO) decree[64] to control the advance of prospecting and fires in the Amazon, the federal government gave the Ministry of Agriculture, Livestock, and Food Supply (MAPA) the right to make decisions on policies aimed at maintaining and monitoring public forests. This is an explicit blow to environmental preservation and regulations won after decades of struggle.

Under the leadership of Tereza Cristina Corrêa da Costa Dias, the Ministry of Agriculture has become closely aligned with agribusiness interests, including beef and soy farmers. Upon taking office, the Bolsonaro administration transferred the responsibility for certifying indigenous territories as protected lands from the National Indian Foundation (FUNAI) to the Ministry of Agriculture. This move, seen as a major victory for the agriculture industry, allows for greater access to land for agricultural use and has been heavily lobbied for by large farmers and ranchers. Tereza Cristina, associated with the 'ruralist' movement, has supported this shift, which threatens to roll back decades of progress made by Brazil's indigenous communities in securing their land rights. The Ministry of Agriculture was also made responsible for overseeing the identification, delimitation, demarcation, and registration of lands traditionally occupied by indigenous peoples, a change that has sparked outrage among environmentalists and Indigenous advocates.

According to Decree No. 10.347 of 13 May 2020, MAPA will now fulfil all government responsibilities related to forests under the terms of Article 49 of Law no. 11.284 of 2006. The requirements of this law include regulating management of public forests for the sustainable production of wood and non-wood products, as well as services related to nature, such as adventure sports. In addition, they include management of the Brazilian Forest Service (SFB) and the National Fund for Forest Development (FNDF), which have provided good policies for the Amazon in the last two decades.

The new government decision has revived the controversy that began at the beginning of Bolsonaro's mandate, when he tried to transfer the duties of the Ministry of the Environment (MMA) to the Department of Agriculture, a measure struck down by the Supreme Federal Court (STF), which understood the decision created a conflict of interests between the sectors involved. Experts and environmentalists warn that combining the forces of MAPA and GLO, which will be coordinated by the National Council of the Amazon (CNA), now under the direction of Vice President Hamilton Mourão, without consulting the Amazonian states and municipalities, could lead to catastrophic consequences for the region.

IBAMA and ICMbio on the front line[65]

The action of IBAMA and ICMBio in Uruará, in the area along the Trans-Amazonian Highway (west of Altamira), 1,021 km from Belém, is another episode in the story of Salles' assembly line of destructive policies. The Uruará case gained notoriety because of the firing of agents responsible for the operation, due to their criticism of the President.

In April, accompanied by just two National Guard officers, one of the IBAMA agents, Divanildo dos Santos Lima (transferred from the operational base of the National Indian Foundation (FUNAI) on the Arara Indigenous Territory in Altamira) flew over the Cachoeira Seca Indigenous Territory by helicopter and located an illegal loggers' camp. According to a statement from IBAMA, published the following day, when the IBAMA team approached, the loggers abandoned four trucks and two tractors. One of the trucks still had the keys in the ignition and was removed from the site while the others were burned. IBAMA agents are empowered to destroy any illegal offices, vehicles, or bridges used in land-grabbing and prospecting operations on Indigenous lands. Article 101 of Decree 6.514 of July 22, 2008, states that upon discovering an environmental infraction, the 'acting agent, in the use of their police powers,' can apprehend and 'destroy or disable products, subproducts, and instruments of the infraction.'

On the night of the operation, on the way to Uruará, Divanildo and the officers were confronted by a group of protesting loggers. A bottle was thrown at the agent, who suffered a head injury. The protest was not unexpected – it had already been prohibited by the State Court of Justice, which had ruled that interventions by loggers in the area were illegal as they risked transmitting COVID-19 to the Indigenous population. The president of the Association of Civil Servants in Environmental Management (ASIBAMA/PA), William Fernandes, declared that 'incidents such as this in Uruará have occurred increasingly frequently since 2017.'[66] And biologist Alex Lacerda de Souza, vice president of the association, affirmed that 'attacks have become more frequent, almost routine, in recent months, creating a troubling scenario, adding greater risk to surveillance agents working in the field.'[67]

The statement from ASIBAMA following the operation called for 'protection and guarantees of safety in surveillance operations and in combating illicit practices that harm the environment.' It also said that IBAMA and the Ministry of the Environment had a duty to guarantee the safety of its field agents, adding that since last year, both the president of IBAMA, Eduardo Fortunato Bim, a Salles appointee, and Minister Salles, have contributed to inhibiting the destruction of goods apprehended in the attempt to prevent illicit activities in the Amazon. There have even been attempts to modify Decree 6.514 of 2008, thus encouraging 'protests against surveillance operations' and the combating of environmental crimes. ASIBAMA also claimed that 'the words of the government serve to justify the acts of violence against federal environmental agents,' and the risk 'agents face in conducting their activities' is high.[68]

In addition to ignoring protests, fires, and physical attacks, the current administration at the Ministry of the Environment has initiated a wave of dismissal proceedings against IBAMA and ICMBio staff when they act against environmental infractions. The discord between IBAMA and its agents persists, encouraging a climate of conflict inside the organization, and providing an excuse for government apologists to feed social networks with memes and fake news decrying the 'negative' actions of IBAMA and ICMbio in protecting the environment of the Amazon.

In the three decades since its foundation IBAMA has never experienced such radical changes as those it has experienced under Salles' administration. On 28 February 2020, for example, 21 of its 27 regional superintendents were suspended. On 14 April, the federal government fired its director for environmental protection, Olivaldi Alves Borges de Azevedo, and replaced him with another army officer, Olimpio Magalhães. ASIBAMA protested at the time that this was in retaliation against IBAMA agents who had tried to combat illegal prospecting on the Apyterewa Indigenous Territory, located in São Félix do Xingu in Pará. The Association also stated that 'the dismantling of environmental surveillance has had a direct impact on the contamination of Indigenous peoples and the advance of the COVID-19 pandemic.'[69]

Armed forces and the Ministry of Agriculture

The government, as always, has reacted on impulse. At the same time as agents from IBAMA and ICMBio were protesting the Uruará[70] case, the federal government published Decree 10.341, authorizing the use of the armed forces 'in enforcing the law and order guarantee decree (GLO)[71] as well as in subsidiary actions along the border, on Indigenous lands, on federal environmental conservation units, and other federal areas in the states of the Legal Amazon.'[72] The GLO decree will be valid for just 30 days (11 May–10 June 2020). It mandates the armed forces to take preventive actions against environmental crimes.

The government's use of the GLO in the Amazon was in reaction to another attack, on an IBAMA agent in Uruará, Pará, on 5 May. The third article of this decree, in particular, is notable as it empowers 'the

Ministry of Defense to determine the allocation of available resources and agents who will be in charge of the operation,' in 'coordination with public safety agencies' and 'public environmental protection entities.' It explicitly states that, while the decree remains in force, 'federal public agencies and environmental protection entities' (IBAMA and ICMBio) will be 'coordinated by the officials referred to in art. 3 of the Decree' – that is, the armed forces. The operation, named Verde Brasil 2 (Green Brazil 2), is focused on combating illegal deforestation and fires in the Amazon, and has been awarded R$60 million to finance 100 vehicles, 20 boats, 12 airplanes, and a force of 4,200, including 400 police officers, according to information provided by Minister Fernando Azevedo e Silva (the Minister of Defense) at a press conference in Brasilia[73]. The military has been tasked with preventing and suppressing environmental crimes through 10 June 2020. General Hamilton Mourão, who heads up the National Council of the Amazon (CNA), declared in a press conference that Amazon Fund[74] resources, which have been frozen, need to be revitalized to finance surveillance and monitoring operations in the region[75]. Mourão also stated that use of GLOs should be made seasonally, to prevent prospectors, loggers, and squatters from returning.

A clear example of the ineffectiveness of the Verde Brasil 2 operation was the action sparked in Nova Ubiratã, Mato Grosso, on 11 May 2020. Two weeks earlier, agents from IBAMA and ICMBio had visited the region and indicated that the area surrounding the Estação Rio Branco Ronuro Conservation Unit, inside the Xingu Indigenous Park, was the focus of illegal[76] activities. But the fruitless foray of the armed forces into Nova Ubiratã, which mobilized 80 agents, two helicopters, and dozens of vehicles, had been against their recommendations. Despite the spectacle – the agents had been accompanied by the media – the operation was declared closed. According to the environmental agency, this occurred 'with no administrative action on the part of IBAMA, given that no request for such procedure had ever been made.'

The government had adopted the same measure the previous year, in August 2019, when fires had raged with unprecedented ferocity in the Amazon Forest, an episode that placed the country in a difficult position on the global stage[77]. During this first use of a GLO decree in the Amazon, R$124.5 million was invested in forces and equipment to contain the advancing fires – in fact, the cost of the GLO was higher than IBAMA's entire budget (R$117 million) for all of 2019. Yet, the measure, which lasted through October of that year, did nothing to impede the advancing fires or the destruction of the forest, as seen by the explosion in the number of prospectors that has been recorded since January 2020.

In April 2020, deforestation alerts increased by 51.45 percent, covering an area of 796 square kilometers, as reported by the National Institute for Space Research (INPE). This marked the highest point since August 2019. Salles told journalists in Brasília that the mission's objective was to stop this percentage from increasing. According to the Minister, the government was aiming for a more consistent reduction in deforestation in 2021, and he added that the GLO had come at a good time, as IBAMA had experienced losses among its staff. These losses of course have been the result deliberate weakening of the agency promoted by Salles and the federal government since the beginning of Bolsonaro's mandate. Salles, however, blamed previous administrations for the problem.

The demands by critics of this administration have continued to grow in direct proportion to the increase in fires, deforestation, and land-grabbing operations, as well as in those processes that are often forgotten by the press, such as fines overturned in legal proceedings typically devoid of transparency. The myriad decrees, official announcements, interviews, and ordinances have added to the sense of chaos surrounding attacks on the environmental integrity of the Amazon, and consequently, on the traditional peoples of the forest. Indeed, the Indigenous inhabitants of the Amazon are the most affected by this destructive chaos, which has opened the door to the spread of COVID-19.

Bolsonaro continues to support prospecting

Traditionally, the Ministry of the Environment handles cases of incursions into the forest and attacks on Indigenous people by loggers and prospectors, although its initiatives tend to have a whiff of marketing spin about them, such as the discourse that emphasizes the 'urgency' of its investigations, or bold statements declaring that 'nobody will intimidate the government'.

Despite these claims, President Bolsonaro continues to support prospectors and loggers. He recently welcomed a major in the army reserves – a former politician from Pará, Sebastião Curió – to the Planalto presidential palace. Now age 81, and confined to a wheelchair, Curió has been heavily involved in repression in the Amazon, from the Araguaia Guerrilla War against the military dictatorship in the 1960s onwards. Curió's repression against the Araguaia guerrillas included torture and killings in which he confessed he had participated[78]. During his career in the Amazon, he spent time in Serra Pelada and ended up, in 2000, as the first mayor of Curionópolis, a municipality born out of a cluster of prospector dwellings.

The content of the meeting between Bolsonaro and Curió was not divulged, but Curió himself is emblematic of the history of prospecting in the Amazon, which is also a history of poverty among the thousands of prospectors who remained in Serra Pelada.

VII. Will the Amazon Rainforest Become a Commodity?[79]

This article from 2018 looks at a decision by the Santarém Municipal Council in Brazil that could be a harbinger of the fate of the Amazon Forest.

Almost two centuries ago, two leading British naturalists and explorers, Henry Walter Bates[80] and Alfred Russel Wallace[81], spent three years studying animals and insects in the region of Lago do Maicá (Maicá Lake) in Santarém municipal district in the Brazilian Amazon. Despite the hardship, the men revelled in what they called the 'glorious forest'. It is estimated that by the time their three years was up, they had collected more than 14,000 species of animals and insects. It was part of a long and demanding trip that eventually led to the publication of Bates's *The Naturalist on the River Amazon* (1863)[82], still regarded as a classic.

We can only imagine what they would have thought of the decision by Santarém Municipal Council in 2018 to change its policies so that part of the lake could be turned into a private port for soybean transshipments. In the Council's last session of the year, 11 December, it hastily and secretively altered the final review of the Participatory Master Plan (PDP), a legal document drawn up in accordance with the city's statutes and approved after a participatory process involving Santarém civil society in November 2017. The aim of the hasty decision was to make it easier for companies to build private port complexes for soybean transshipments on Lago do Maicá. The decision invalidates months of discussions that included working groups and audiences with representatives from the most diverse sectors – the business community, academics, public entities, social organizations – and it violates Convention 169[83] of the International Labor Organization (ILO). The holding of the special plenary session came as a complete surprise to traditional populations, inhabitants, and social movements in Santarém and Amazônia.

The large new port will be built in the Lago do Maicá region, an area of great environmental complexity, inhabited by traditional communities, fishing communities, and around 400 Quilombola families[84]. Altogether, about 1,500 families live in the area. Maicá is an ecological sanctuary, a natural breeding area for unique species of aquatic fauna and Amazonian bird life. Besides being a tourist attraction, the lake is also a source of income for the families who live primarily off fishing, and provides about 30 per cent of the fish[85] consumed in the town.

Archaeologist Anne Rapp Py-Daniel of West Pará Federal University (UFOPA) told me in 2018 that she is worried about the impact of the passage of large ships on the lake:

> *Lago do Maicá is an extremely rich, but also very fragile, ecosystem with a large-scale geological formation dynamic [land fall, land in formation, drill holes, etc.]. The presence of large ships will cause significant water displacement and alter the dynamics of the river currents, leading to accelerated destruction of the lower wetlands where many traditional communities live. Monitoring has already been carried out on the Madeira River, another wetland region, showing the harmful impact of ships and barges.*

Furthermore, the Maicá region is extremely important for archaeology, as it houses the oldest known archaeological site in the municipality, Sambaqui de Taperinha, which is 8,000 years old. We also have a large number of more recent sites between 2,000 and 500 years old that are still being mapped, many of which are identified by traces of *terra preta* (dark, fertile, anthropogenic soil created by Amazonian Indigenous communities). Indigenous communities are also living in the higher areas. The history of this region doesn't stop there: Quilombola settlements have been in the area since the 19th century, with the addition of nine territories recognized by the Palmares Cultural Foundation along the banks of Maicá/Ituqui.

Centuries of history

Santarém is one of the oldest cities in inland Amazônia. Located at the confluence of the Tapajós and Amazonas Rivers, it was founded by Jesuit priests in 1661, when the Portuguese colonized the region. Since then, Santarém has been a strategic production center, beginning with the production of cacao, and then livestock, rubber, jute, and currently the soybean monoculture. Located 475 miles from the Atlantic Ocean, its geographical position is strategic for the transport of soybeans, whether the crop arrives by the BR-163 road or by barge on the Tapajós. It then travels along the Amazon River to the Atlantic.

Construction of the port zone in the Lago do Maicá region is part of the strategy of the region's soybean farmers and trading companies to expedite the flow of grain from Mato Grosso up to northern Brazil, more precisely, through the Tapajós–Teles Pires axis. Pedro Martins, from human rights organization Terra de Direitos, observes:

> *Soybean farmers appear to have started a process of usurping the lands from the rural dwellers. These farmers, generally coming in from other states, have begun to plant soybeans in large plantations in the region of the Santarém Plateau. This is how Embraps[86] came to be, due to the arrival of soybean farmers from the Mato Grosso region who wanted to facilitate soybean exports through Santarém, but who also see enormous profit potential in port construction in the region.[87]*

Lago do Maicá hydrographic basin

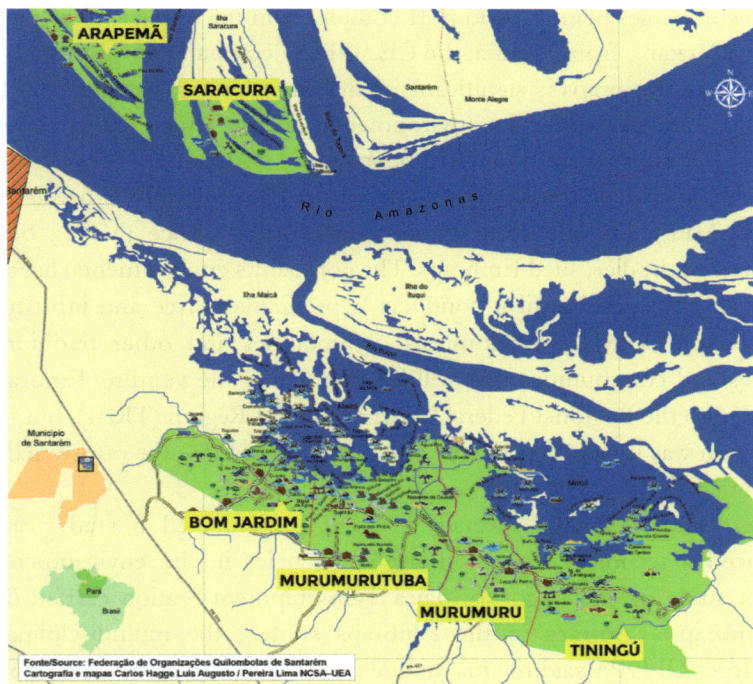

Lake Maicá hydrographic basin

For Amazonian priest and activist, Edilberto Sena, the process began with the decision by Cargill, a multinational company, to build a terminal in 1999:

It saw Santarém as a strategic place to lower the cost of soybean exports from Brazil's Central-West region. Local politicians and even some members of society mistakenly believed that the multinational's port would bring jobs, income, and development. But this was a trap for the population, for all it did was seriously harm Santarém society. People living on the outskirts of the city – Pérola do Maicá, Área Verde, Jaderlândia, Jutaí, and five other areas – are now going to have to deal with a massive new highway. Because it will have the capacity to handle 800 trucks a day, you can imagine the number of accidents and other problems that will be created. If these populations don't get organized, if we are not together with them in resistance, the destruction of our city will get worse, because the authorities have no respect for human life. This Embraps port may be useful for business but will cause serious harm to the environment and to the inhabitants of Santarém, as has already been the case with the Cargill port.[88]

Three companies plan to build port complexes in the municipal district – Grupo Cevitai[89], from Algeria, the CEAGRO[90] company, and Embraps[91]. The case of Embraps is significant, as the environment secretary of the State of Pará suspended its environmental license after a court action. The lawsuit arose when traditional peoples and communities living in the Lago do Maicá region, with the support of the Federal Public Prosecutor's Office (MPF) and the State Public Prosecutor's Office (MPE), both independent bodies, sued Embraps. The company's environmental license has been suspended until it conducts 'a preliminary, free and informed consultation' with the Quilombola communities and other traditional peoples and communities that will be affected by the venture. Embraps appealed to the Regional Federal Court of the 1st Region (TRF1), but the injunction was upheld and the environmental license for the port remains suspended.

West Pará Federal University (UFOPA) conducted a study[92] and produced a technical report on the deficiencies in the environmental impact study of the Lago do Maicá Transshipment Station carried out by Embraps. Contrary to the Embraps study[93], the multidisciplinary team from UFOPA said the magnitude of the environmental and human damage would be such that, if not remedied, the fish and phytoplankton populations of Lago do Maicá would be at risk, and there would be irreversible damage to human and nonhuman life in the region.

Jackson Rêgo Matos, a lecturer at the Institute of Biodiversity and Forests (IBEF/UFOPA), also has environmental concerns:

Our greatest concern about construction of the port is not only its impact on Lago Maicá, but what it will mean for the entire city of Santarém and the region of the Tapajós River as well. The landscapes of these areas, including the Alter do Chão Beach, will be negatively affected by traffic congestion caused by trucks circulating throughout the city, as well as by convoys of barges on the river. This logistical movement will certainly mean more air, visual, and noise pollution, as well as the loss of archaeological heritage, as Santarém is the oldest pre-colonial city in Brazil, housing one of the most significant archaeological sites in the Americas. It should also be emphasized that the Tapajós Basin[94] is the fifth largest tributary basin in the Amazon and covers approximately 492,000 km². This alone means that we have to have public policies that guarantee maintenance of this unique heritage for the enjoyment of the population. The mayor, Nélio Aguiar, says the policies will create a source of revenue for businesses, as if that were enough to justify the disrespect council members are showing for the constitutional formalities of the master plan built with citizen participation. But this is not enough.[95]

Autonomous voice

As was pointed out in a report on the *Brasil de Fato*[96] website, the way of life of communities living in the region is being endangered to achieve something they do not need – a shorter exit route for Brazilian soybeans. With construction of the port, the distance required for road transport from Mato Grosso could be reduced by around 800 km (currently most of the crops are taken south to the port of Santos). If the crops were brought to the port of Maicá in Santarém, the journey by ship to Europe would be shortened by a week.

However, according to Mário Pantoja, Quilombola leader in the area of Lago do Maicá, the voice of the local populations needs to be heard:

The main beneficiaries of the port construction are precisely big business. People speak as if we are impeding progress, whereas we are really helping to promote sustainable development and progress. Why? Because we work with fishing. And once the port is built, there'll be no more fishing.[97]

This talk of progress is outdated in any case, says environmental activist, Father Guilherme Cardona. He points out that 'this model of development is creating unsustainable cities just when today's dynamic, all over the world, is how to create sustainable cities so the population and development can go hand in hand.'[98]

Despite being presented as projects to develop the region, the ports will negatively affect nine of the city's districts, inhabited by traditional communities, who now live in an urban area because public policies failed to protect them, and they were evicted from their old homes. Dona Sebastiana[99], who fishes in the Lago do Maicá, says: 'Nobody agrees with this [the construction of the port]. Because we need the lake. Because soon we won't have any more fish to catch, because the land here will be reclaimed, and the fish will disappear.' The same point was made by Quilombola member João Lira: 'The question is: why a port in the Maicá area? Who does it benefit? The people of the region? I believe it brings no benefits, zero benefits, to them.'[100]

Father Edilberto Sena believes the future of the Amazon region needs to be considered within the framework of the environmental, social, and cultural threats it faces from the virulent aggression of outside capital:

> *The dispute for territory – land, forest, rivers, subsoil, and people – is becoming more and more aggressive. Our region, in western Pará state, is an example of what is happening all over Amazônia. We have a 70,000 ha soybean crop invasion, with extensive use of pesticides. We have an invasion of grain ports for soybean exports from Mato Grosso. There are 23 ports either built or under construction on the Tapajós River, a 930 km railway being planned to run between Cuiabá and Miritituba on the Tapajós River, and seven hydroelectric dams planned for the Tapajós River. Finally, the city of Santarém, a center for all this exploration, is being taken up by 20 warehouses, which will force inhabitants out of the downtown area to the periphery.*[101]

How long will economic power corrupt the executive and legislative branches in Santarém so international laws are broken on behalf of an insignificant number of shareholders who do not respect the Amazonian biomes or its people? How long will the Amazon rainforest be at the mercy of council members, politicians, and business people who formulate laws in the dead of night that will harm society and assault communities, traditional peoples, and the environment, and who impose the toxic monoculture of soybeans, with the excuse of boosting the national trade surplus?

We can be sure that Bates and Wallace would not have approved of the decision made by the Santarém City Council. After spending 11 years documenting the beauty of the tropics, Bates said: 'In the end, I was obliged to conclude that contemplation of nature is not enough for human minds and hearts'. Or, as Brazilian writer Euclides da Cunha would later say when describing the region during his visit in 1905, 'The Amazon is the last page of Genesis yet to be written'[102]. If it depended on Santarém's city council members, all that would remain of the last havens of the world's biodiversity will soon be found only on bookshelves, in the records of the natural science books of Bates and Wallace.

Part 2

The Amazon and the Pandemic

VIII. Hunger in the Amazon: The Invisible Companion of COVID-19

This chapter combines two articles: the first[103] was part of an exploration into the increase in food insecurity in the Amazon, and was the result of a research trip taken in March 2020, just as Peru was implementing a lockdown at the beginning of the pandemic. The second[104], examined how the increased dependence on commercial goods by the Amazon's riverine communities, and their subsequent abandonment by the Brazilian state during the pandemic, has led to a twin epidemic of hunger and the virus. Images taken by the author during the trip (see below) were mounted as an exhibition, 'Amazonia Hunger', at Florida State University.

I will never forget the conversation I had with Rafael, a fisherman, from the banks of the Solimões River in the small, isolated town of Tabatinga, Brazil[105]. His words still haunt me: 'In the Amazon, if COVID-19 doesn't kill people, hunger will.'

Reaching the 'heart' of the Amazon

It is less than 24 hours since I arrived home from what I consider to be the ecological heart, the very soul of the planet: the Amazon. I had left Tallahassee, Florida on 12 March 2020 for a research trip to Iquitos, Peru. When I left, the coronavirus was already making headlines around the world, but I decided to go – cautiously – as I would otherwise have had to forfeit all the money I had saved for the trip. My goal was to finish the documentary I was making about environmental issues in the Amazon in collaboration with several local poets. However, a turn of events took me by surprise and forced me into a situation that allowed me to get an even clearer picture of the challenge we face as a species.

Before the circumstances surrounding COVID-19 escalated, I was able to visit several riverine communities around Iquitos and experience the bustle of the former Peruvian rubber capital's famous Belén market. But on Sunday 15 March, at around 8 p.m., Peru's President Martin Vizcarra declared a state of emergency and closed all the borders. He gave tourists less than 24 hours to leave the country. Many foreigners were unable to find a way out and were stranded, without the slightest idea of when they would be given permission to leave.

On Monday 16 March, I tried to catch a flight to the small Peruvian town of Caballococha, in order to continue my journey onward from there to Tabatinga, Brazil, by boat. Brazilian cameraman Bruno Erlan and I arrived at the airport ahead of time with pre-booked tickets. Unfortunately, due to poor weather, the flight was cancelled. We were told there was a boat scheduled to leave for Tabatinga at 7 p.m. We ran to the pier through the chaotic streets of Iquitos. Local police had closed the main roads so traffic was even more manic than usual. We arrived at the pier before 7 p.m., but the boat was already full. There were more than 600 people crammed onto a boat with a capacity of perhaps 300.

I decided to stay in Iquitos and, because I also hold a Brazilian passport (my mother is from Brazil and my father is from the US), I contacted the Brazilian consulate in the city. They managed to negotiate our departure with the local authorities. We were to embark on a cargo boat headed for the town of Tabatinga, situated near Brazil's border with Colombia and Peru. Three days later, we went to the consulate in Iquitos for a medical check-up before leaving the country. Four doctors from Peru arrived in an ambulance equipped to carry out several tests on our group (consisting of a total of eight Brazilians). When this was done, the staff from the consulate escorted us to the boat.

Food supply by boat

We spent three whole days travelling the Upper Amazon. The cargo boat (privately owned) was loaded with four tons of food supplies. These were to be distributed to communities along the way. We stopped in the small 'towns' of Pebas, Nuevo Pebas, Cochiquinas, Alto Monte, San Isidro, San Pablo, Caballococha, and Santa Rosa. The Bora, Huitoto, Tikuna, and other Indigenous peoples inhabit these riverine communities. In all of them, I was able to witness an Amazon that had previously been invisible to me.

Many of these families do not have access to land ownership, nor to the means for sustainable agriculture. Nevertheless, they usually continue to produce some of their food while relying on the boats for staples like sugar, salt, and canned foods, and sometimes paying or bartering their local products for these imported goods.

During all of these stops, I could not help but notice how these Amazonian communities had changed, and how these changes were affecting their livelihoods, culture, way of life, and interactions. They had become almost entirely dependent on supplies that arrived via cargo boats, like the one I was travelling in. The boat provided everything from rice, beans, flour, eggs, water, soft drinks, and all kinds of fruits and vegetables to building materials, tiles, furniture, clothes, etc.

The riverine communities greet the boat as the purveyor of an abundant supply of everything the forest can no longer give them. One of the captains remarked: 'I've been doing this for 30 years and these communities aren't what they used to be. Now you see the river, but that's it; there aren't any fish.' He continued: 'People here don't like farming anymore; they are increasingly dependent on the cargo boat.' For centuries, riverine communities have relied on a similar method of provisioning by the regatão.

Seasonality and migration

Faced with the pandemic, traditional Amazonian populations require special attention. It is well known that there are seasonal shortages of regional staples – products like eggs, turtles, watermelons, and beach beans do better when the waters recede. Also, the international migration of political refugees and those fleeing violence alters traditional riverine relations, introducing a variety of nutritional needs and eating habits. The displacement of landless peoples from the south and southwest to the Amazon emphasizes this change; these populations do not know how to plant and harvest crops in the forest unless it has previously been burned and cleared.

Seasonality also changes the demographic profile of the local population, causing migration from the settlements, and the situation of refugees from both Venezuela and Haiti, as well as displaced populations from various parts of the Amazon, is aggravated by COVID-19. Changes in the way of life of traditional populations have made them dependent on goods from the big cities, arriving in weekly shipments via the river. This means the imposition of controls on movement during the pandemic could be catastrophic.

Market relations in the deep Amazon, a region with little outside influence, explain this dependence on outside goods. These relations limit riverine communities' traditional hunting, gathering, and production activities to family agriculture, growing food under the direction of agronomists. The population of the Amazon is growing exponentially, so how can this limited production meet their needs? How can the provision of basic necessities be reinvented without harming the environment? This is the age-old question at the heart of the problem. Basically, riverine communities of the upper Amazon depend on supplies transported by cargo boats because they are unable to produce their own food. Moreover, they have no access to land ownership, nor the means for productive agriculture, and the practice of gathering food has been affected by controls on land use. Thus, only the most isolated Indigenous fishing communities are able to maintain their subsistence way of life.

The untouchability of nature

'Slash and burn', the traditional method of preparing the land, has been prohibited since the implementation of the Forest Code (Código Florestal) in 2012. The ban is the result of apparently naïve environmental legislation. To put it simply, influential environmentalists from outside the Amazon have defended the idea that nature in the region should remain untouched and undisturbed, arguing that the Amazon should be depopulated in order to preserve it. This has been a recurring debate among 'multilateral' groups since 1992. Some environmentalists claim the traditional technique of clearing and burning the land for planting crops damages the environment, maintaining that burning produces excess heat, as seen recently during the Amazon summers. Hunting, which is seasonal, is also impossible in the period between harvests due to flooding, and current Brazilian legal restrictions are in place at municipal, state, and federal levels. Not only is the value of fish now determined by commercial relations between businessmen in the fishing industry, with no involvement of the fishermen themselves, but it also appears that any fish caught are often not worth selling, so are salted and kept for the off-season.

All this has an impact on land-use changes, as riverine peoples and the Indigenous depart from traditional customs and local food staples to cultivate food they do not usually eat. The expansion of oil crops, grown for export in the southern part of Amazonas State, Humaitá and Porto Velho, is an example of this. Such practices have become increasingly common in the Amazon.

As a result, the acquisition of new food products – primarily, chicken, pasta, beans, dried salted meat, and rice – leads to dependence on wholesale and retail trade, in addition to increased boat traffic. River-dwellers buy these goods in exchange for rare seasonal fruits, fresh and salted fish, and even game. This 'system' of dependence actually has a long history, at least as old as the rubber trade, which today has been replaced by logging and other forms of extractivism. Dependence is nothing new, but it is increasing with the availability of more regularly supplied goods. In other words, traditional activities are being replaced with market relations no longer dependent on nature but on trade routes. Dependence on commercial goods (processed and junk foods) also means greater waste in the form of plastic and cans that pollute the water and wash up on the shore.

Yet, the Indigenous peoples of the Amazon are endowed with extraordinary knowledge that is invaluable to humanity. Now more than ever, we must listen to them and preserve their cultural and ecological values so as to ensure their survival and, consequently, the survival of the Amazon. Otherwise, colonization and dispossession of traditional customs will continue, with the spectre of COVID-19 only serving to accelerate the process.

Exhibition: Amazonia Hunger

In March 2020, after a challenging journey from Iquitos, Peru, to Tabatinga, Brazil, and back to the USA, this digital exhibition was created to shed light on the critical issues faced by the Amazon's remote communities during the COVID-19 pandemic. Developed by Florida State University's Office of Digital Research and Scholarship, it captures the experiences of various tight-knit ancestral communities living along the vast and serpentine rivers of the Amazon. As the pandemic swept through South America, these isolated populations confronted unique challenges exacerbated by their reliance on external goods and disrupted supply chains. Launched in March 2020, the exhibition is still available online[106], offering a vivid portrayal of the profound impacts of the pandemic on these vulnerable communities.

'Amazonia Hunger'

The labyrinth of life

After our second day of filming in Peru's Iquitos region, we went down to the Ananais River. We took out our drone and contemplated the scenery – the river's serpentine curves reveal the beauty of the Brazilian-Peruvian Amazon. But the winding river and vast forest seen here are both under threat. The river has served as a battleground for pro-mining efforts that have contaminated the waters, and the forest is slowly shrinking.

A city on the river's banks

On the banks of the Itaya River, just across from the larger city of Iquitos, live the river dwellers, in their floating houses. In the city, the historic Malecon Boardwalk of Iquitos sits at the very edge of the water. This panoramic view depicts the contrast between urban and rural, yet both rely on the surrounding environment.

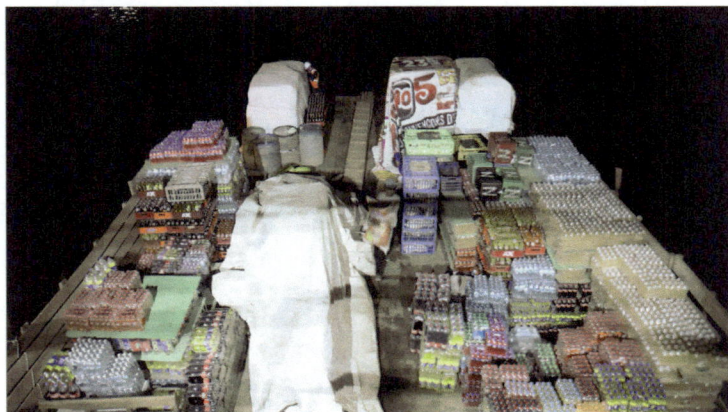

A forest diet?

Taken minutes after leaving Iquitos, this photograph shows the bow of the cargo ship, Maria Fernanda, with the lights of the city reflected in the water. Before we could board, we were sent to several health service stations, and a group of doctors came to the consulate to check us for any symptoms COVID-19.

The sunset's kiss

After we departed Iquitos, we stopped at several communities on our journey to Tabatinga. On board the ship, we watched as the low sun and pastel skies were reflected in the waters of the Amazon, softening the view, and enticing us to gaze at the overwhelmingly beautiful sight of the forest.

Awaiting the tide

Several women were standing waiting for the cargo boat, their heads veiled. Most of them belonged to a strict religious group that lives in a world that is vastly different and more challenging than that of other groups in the Amazon.

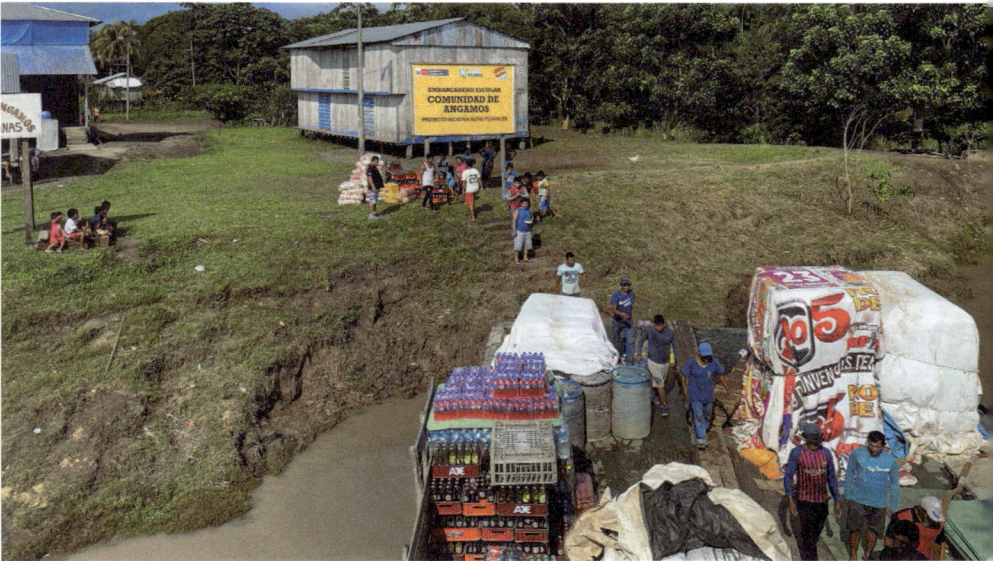

Waiting to unload

Many members of the Angamos community were also standing waiting for the essential supplies, including rice, beverages, and raw materials carried by the cargo boat.

River GPS

This panoramic view of Pebas community shows it reaches deep into the forest, where many people live far from the dense urban areas. All of these smaller communities have ports on the Amazon River, through which they access goods made in the larger cities.

Forest architecture

Rather than skyscrapers and apartment buildings, the architectural landscape of the Amazon is filled with wooden stilt houses, built to avoid the rising river tides. The families here depend on the river for their livelihood.

Canoe vision

When the COVID-19 virus first made global news, the Peruvian borders were closed. We were forced to leave the country on the Maria Fernanda, traveling from Iquitos, Peru, to Tabatinga, Brazil. This photo is an aerial view of the boat, which stopped in about 10 different communities to bring food and supplies to communities in need.

Reframing life

As the sun sets behind the community, the window frames a scene depicting the solemn beauty of the life of these river dwellers. Although we were inside the boat, cut off from it, the moment nevertheless offers a glimpse of the routine and life of the Amazon.

Playtime in the jungle

The boat stopped in the Cochiquinas community, where we saw these children playing at the river's edge while waiting for their families who were helping to unload cargo.

A bend in the river

A panoramic view of the Amazon River near one of the local communities on the way to Tabatinga. On one of our final stops of the day, we photographed this sunset at the edge of town, which is just one of many that are entirely dependent on the weekly supplies brought from the city.

Amazonia's nibbles
One of the workers on the boat, hired to unload the cargo, is caught here stopping for a quick break to enjoy a coconut.

Open for business
The Pebas community comprises many families who live in homes built out over the water. They receive their supplies through this family. The young man here has filled his canoe with crates of fruit and vegetables to be delivered to other members of his community.

Waterside exchanges

A woman is seen trying to sell her produce to potential passengers on the boat. However, due to COVID-19, the only people on board were the crew members and us. The boat was already nearly empty, but one of the young men is shown here unloading a sack of onions.

Morning commute

People from the Fila Del community leave to go fishing, one of their daily routines. This inconsequential moment (for them) seemed meaningful to us – instead of cars or bicycles, they use boats to commute to work, and their daily routes are the rivers.

Supply and demand

This was yet another town, toward the end of the day. We had stopped in various communities throughout the day, where residents eagerly waited for the boat to bring what they needed. With each stop, the provisions on board diminished.

Papaya for sale

A local woman came to sell papayas to the boat's passengers, but this time there was no one but us. We ended up buying some of the delicious papayas this especially kind woman offered for sale.

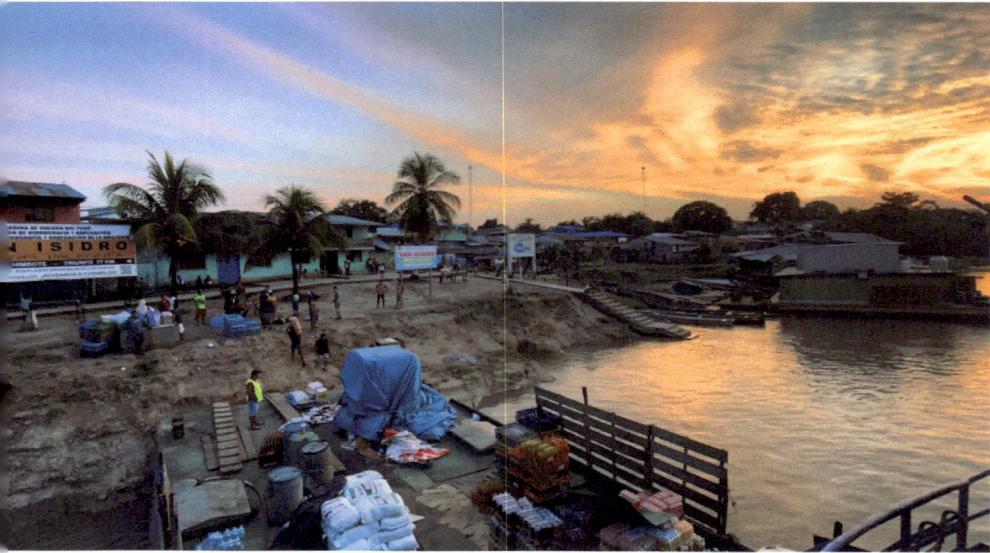

Sunset in San Isidro

In this photo, the sun was setting over the riverside community of San Isidro, where residents came to unload goods from the boat. The cargo was slowly dwindling in size as our journey from Iquitos to Tabatinga neared its end.

Material life

This was yet another town, towards the end of the day. We stopped in a series of communities during the day, in all of which residents eagerly waited for the ship to bring what they needed.

Amazônia parking lot

In the early days of the lockdown in Iquitos, I watched as empty boats gathered, with increasing numbers of people concerned about this sudden emptiness. The true danger for many in the Amazon was not the pandemic, but the hunger that would come with it.

Homemade food I

This man was attempting to sell his fish at the riverside, even after Colombian and Peruvian borders with Brazil had closed.

Houseboats

Groups of houses like this, sitting directly on the Amazon River, are found all over the region, as many individuals rely on the river's resources for their livelihood.

Homemade food II

In the municipality of Caballococha, family members eagerly helped their father, who had arrived with his catch of the day. His canoe was pulled up in front of their stilt house.

Fisherman's apprentice

While traveling along the Amazon River, we found this boy in Iquitos cleaning fish caught by his father to eventually sell. We came across the boy as we were trying to find shots that captured the essence of the city and struggles of the locals. At this point, the residents were still managing to sell their products and survive, despite beginning to feel the consequences of the border closures caused by the pandemic.

No soccer today

In an important area in the Belén region, this photograph displays several of the details of life in the Amazon. There is a school entrance on the left, while the soccer field is in the middle of the plaza surrounded by houses. Trash in the rising water is proof of the growing problem of pollution.

Last stop
Belén District is one of 13 districts in Maynas Province and it is well
known for its market, which sells anything caught and raised in the region,
drawing fisherman and farmers alike to the city. We were able to ride
through the entire Belen market by motorcycle, but stopped at the flood
plains at the edge of the Itaya River where this photograph was taken.
The city is home to around 60,000 people, most of whom live in extreme
poverty along the banks of the river.

Cooling down in the sun
As we passed by the river's edge in a canoe, we saw this man alongside his
house, refreshing himself on a hot day.

River front yard
This picture shows a local stilt house, or palafita, that sits on the Amazon River and is built to protect the family from flooding. Their boat, loaded with local resources like wood and fish, sits outside the house, and serves as their main means of maintaining their livelihood.

Family photo
We had the opportunity to spend some time talking with this family, who lives in the palafita house in the previous photo.

Lunchtime by the river

We were filming on the Peru side of Tabatinga when we came across this girl and her mother eating their lunch, laid out on banana leaves.

Silent tide

Sunset, after a long day of filming on the Ananais River. This fisherman's hut is a typical house built on one of the many rivers of the Amazon.

Arriving home

In downtown Iquitos, we spotted this local fisherman listening to music in his house at the river's edge. When asked how he thought things were going regarding COVID-19, he replied that he was concerned, but not afraid.

Coinless laundry

After a long day of filming on the river, we returned to Iquitos where we encountered one of the river dwellers washing his clothes outside his house. This typical Amazonian scene shows the daily lifestyle of the ribeirinhos.

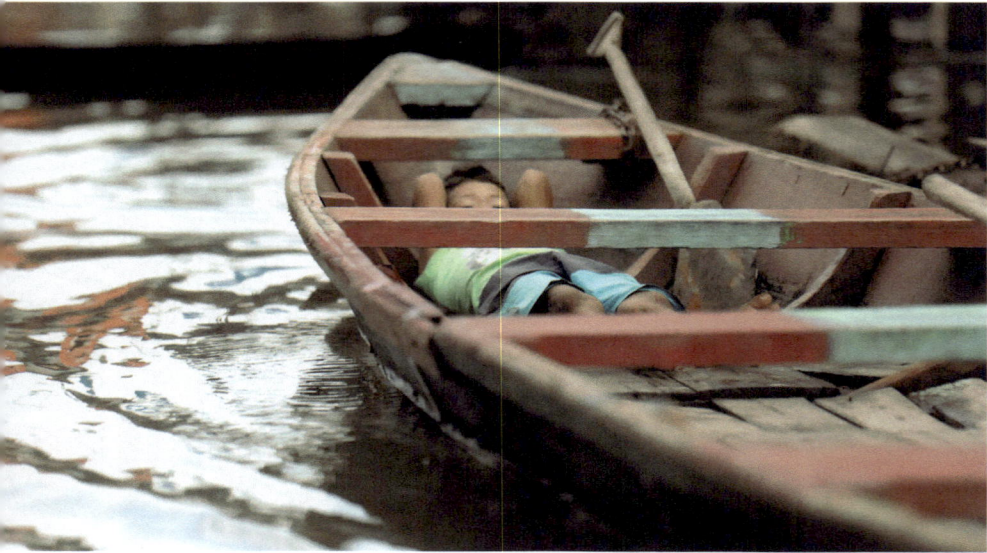

Canoe hide and seek

A young child hid himself in the canoe as he was about to go fishing with his father. His playfulness subtly touches on how children in the Amazon live their daily lives.

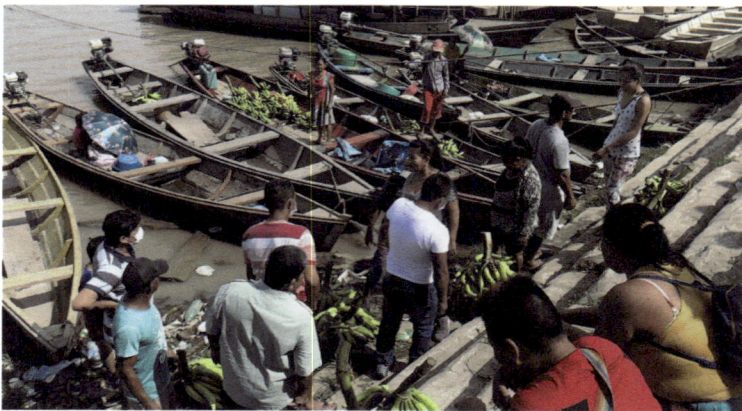

Amazônia marketplace

The cargo boat finally arrived in Tabatinga, where we spent five days waiting for our plane. Although the borders were closed, we saw many Peruvians who had crossed the border into Brazil to sell their bananas and other products in Tabatinga out of necessity, because they could not otherwise afford to live.

Hands on

A local family in Iquitos preparing a Piracatinga fish for their dinner. While the fish is considered a delicacy in the region, in other regions, it is thought to be unhealthy to eat. These regional fish are an important part of the Amazonian diet, and many families depend on the river for their food.

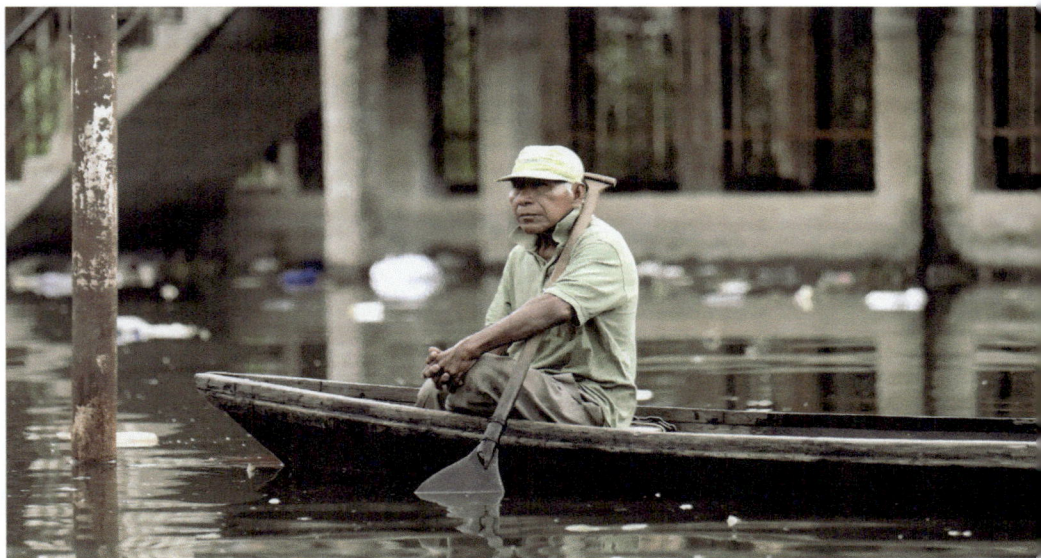

Hands of hope

An older man, whose face shows his experience and knowledge of the local fishing scene, looks sternly at the river ahead of him. In the background, trash floats along the surface, but it seems as if his search for hope is distracting him from his own circumstances.

No way to get lost
This picture, taken on the Solimões River, captures the beauty and magic of the Amazon as the river bends and flows, cutting through the forest, which is intensely green under the bright sky.

IX. Deregulation and Deforestation Fuel the Pandemic in the Amazon[109]

In this 2020 essay, the author argues that acceleration of Amazonian deforestation and environmental degradation under Bolsonaro has amplified the lethality of diseases, especially COVID-19, in a region where public healthcare is at best precarious.

Removing protection from Indigenous lands

In 2020, the world watched with bated breath as Brazil saw a five-fold increase in the number of deaths from COVID-19 in less than a month. In the same period, the country's environmental protection policy also suffered at least five lethal blows.

On April 22, 2020, exactly 520 years after the occurrence of the largest genocide in the history of Brazil, the so-called 'discovery' of the country, the National Indigenous Foundation (FUNAI), the agency charged with promoting the defense of native people's rights and guaranteeing the health and integrity of the territories occupied by these groups, published Normative Instruction (Instrução Normativa NI) 09/2020. This states that all Indigenous lands not in the final stage of state recognition will be excluded from the database of the National Land Management System (SIGEF). In practice, this NI removes protection from the vast majority of Indigenous territories, including many that have spent long years in claims procedures, and those occupied by groups of uncontacted Indigenous peoples . It validates private property holdings and title deeds previously annulled by Article 231 of the 1988 Brazilian Constitution, and it encourages non-Indigenous occupation of these lands. It is crucial to remember that Indigenous lands are the areas with the country's lowest rates of environmental degradation, both because they themselves are conservationists by practice, tradition and values and because they are on the frontline, defending their lands against those who arrive to degrade and destroy them.

On April 30, two of the main environmental surveillance coordinators at the Brazilian Institute of the Environment and Renewable Natural Resources (IBAMA) were summarily removed from office. Everything indicates that these firings were in retaliation for their work investigating and combating complex grandiose projects that pillage natural resources. The dismissals occurred soon after a large-scale operation to combat illegal mining in the Amazon, where the team, led by the two civil servants, had managed to reduce deforestation in 2020 to zero in the Ituna–Itatá Indigenous Territory, until then the worst-affected area. This was even more the more remarkable given the severe official constraints already limiting IBAMA's operations.

Less than a week later, President Bolsonaro published Decree No. 10,344, placing all federal environmental protection agencies and organizations under the command of the armed forces. The Chico Mendes Institute for Biodiversity Conservation (ICMBio), IBAMA, the federal police, and FUNAI have all lost their operational independence and are now subordinate to the military, who have the power to determine the locations to be inspected and the procedures to be adopted.

The same act also authorized a Law and Order Guarantee decree (GLO) in the Legal Amazon area May 11–June 10, 2020. This GLO is estimated to cost around R$60 million (about £9.6 million). IBAMA's environmental surveillance budget for the entire year is R$74 million (£7.9 million). In other words, more than 80 per cent of the year's available resources will be used just on implementing the GLO decree.

This money has primarily been used to deploy the Brazilian Armed Forces, supposedly to support local law enforcement agencies in combating illegal activities such as deforestation, illegal logging, and land invasions. Expenditures under GLO include personnel costs, logistics and operational expenses, infrastructure support, surveillance and monitoring technologies, and training programs.

Yet, since introduction of GLO, there have been no reports of confiscation and destruction of even a single machine used in illicit activities. It is important to remember that the destruction of backhoe excavators, crawler tractors, and other machines, each costing R$500,000 (£80,000) on average, is an indispensable legal mechanism that gives teeth to surveillance operations. In the past, these types of operations were conducted mainly by IBAMA, and they caused significant pain to its opponents, many of whom occupy important positions in state legislatures.

On 12 May the precarious state of the already debilitated ICMBio worsened. The agency is responsible for management and surveillance of federal conservation units, areas of great environmental importance. The decree abolished 11 regional coordination centers, replacing them with five regional administrations for the entire country, four of which are situated in cities far from the places they are supposed to protect. Previously, there were five regional coordination centers in the Amazon. Now, there is one regional administration for the entire biome, which contains around 130 conservation units. Most of those appointed to the new administrations are retired military personnel, with no previous experience in environmental management. These appointments further swell the military ranks of the agency, whose presidency and four general-director posts are occupied by military police officials from São Paulo. It is no coincidence that the position of surveillance coordinator remains vacant.

This aberration is typical of the 'necropolitics' now prevailing in Brazil. It constitutes a clear attempt to subordinate all public agencies and institutions to the whim of the executive branch, in a direct challenge to constitutional power and authority. This is one of the reasons why the V-Dem Institute of Gothenburg University drastically demoted Brazil in its ranking of states on the democracy-authoritarian scale.[110]

A bonfire of forest regulations

The government and its supporters, however, are not satisfied with their attacks on agency surveillance and the regulation of illegal land grabs. They have turned their attention to redefining the types of deforestation that can be authorized. The measure they have used is Decree 10,347, published 14 May which, in direct contradiction to a provision of the law that regulates the administration of public forests, excludes the technical division of the Ministry of the Environment from determining areas where forest products can be exploited. The legality of this decree is being challenged in court through an *actio popularis*[111] in Belém.

Under this new decree, the drawing up and approval of the Annual Forestry Grant Plan (PAOF) becomes exclusively the job of the Ministry of Agriculture, Livestock, and Supply (MAPA), whose overriding function is expansion of the agricultural frontier. To get an idea of how catastrophic this measure may prove, the PAOF plan for 2020 envisages making 7,750 million hectares of federal public forest subject to concession. This represents an area 50 times the size of the city of São Paulo, and eight times larger than the total amount of clear-cut deforestation detected by satellite August 2018–July 2019, which in turn was at the highest level seen in a decade.

Strictly speaking, forestry concessions authorize selective felling of trees for timber, which in theory should reduce the scale of forest loss. However, the reality is more concerning. Timber extraction almost always culls all forestry species that have any market value and ignores the felling cycles of proper forest management. The plans filed are almost always fictitious, serving to cover the looting of wood from Indigenous lands and special protected areas.

On top of all that, the question arises: why does a ministry, whose purpose is to promote rural activities that typically require vast open areas, such as agriculture and livestock, want to monopolize the fate of public forests, even if, in theory, they are to be exploited through selective logging? In 2020, this affected almost 8 million hectares. From 2021 on, there was no guarantee that this number would not increase significantly. The scenario is frightening: uncontrolled illegal deforestation breaking new records each year, and deforestation authorized by those whose job is to open up new areas for agriculture. This is a recipe for disaster.

Clearing the ground for pandemics

According to the World Health Organization (WHO), pandemic is the term used to describe a disease that quickly and simultaneously spreads to various parts of the planet. The same term might be used to define the virulent spread of large-scale agribusiness in the various biomes of the world.

It is also crucial to recall that destruction of the forests is directly linked to the appearance and escalation of diseases. As Carlos Nobre emphasizes: 'The Amazon has the highest number of micro-organisms in the world. And we are disturbing the system all the time, with urban populations getting closer, deforestation, and the trade in wild animals.'[112] In today's dystopian scenario, Amazonian peoples find themselves in even greater danger than the forest itself, given their multiple vulnerabilities, the losses they suffer, and the biological chaos that threatens them. If the forest is cleared, a valuable system protecting human health will also be lost, not only for those who live close to the forest, but for the entire nation. The forest provides a health service of inestimable value when it is the source of quality water, food safety, and climate equilibrium. Deforestation and other forms of environmental degradation are a direct cause for the appearance of various diseases with a large-scale social impact, both locally and globally.

The reduction of Amazon Forest cover transforms the naturally acid pH of the forest into a pH close to neutral, creating favorable conditions for proliferation of the Anopheles mosquito, the vector of malaria. The accelerated deforestation that began in the Amazon in the late 1970s brought with it a proportional increase in cases of malaria, reaching a peak in 1999, when 632,000 cases of the disease were recorded. After 2005, malaria cases decreased almost 80 per cent through a series of diagnosis and treatment measures, falling to 130,000 cases in 2016. Sadly, but predictably, as deforestation in the last two years has intensified, the number of malaria cases has once again increased by more than 50 per cent.

In 2016, Amazonas was also the state reporting the highest incidence of tuberculosis in the country: 67.2 cases per 100,000 inhabitants, more than double the national average.

The poverty, inequality, and precarious sanitation prevalent in most Amazonian cities, resulting from decades of predatory development, provide excellent conditions for the proliferation of diseases such as dengue, chikungunya, and Zika, which were previously more common in other regions of Brazil. Currently, the Amazon is responsible for 95 per cent of the cases of Chagas disease in Brazil.

Shattering the equilibrium between society and the environment

When the fragile equilibrium between the environment and society is broken, it leads to the appearance of new diseases and the reappearance of forgotten evils. It is no accident that the five cities with the highest rates of COVID-19 lethality are located in the Amazon region (Tabatinga, Manacapuru, Autazes, Coari, and Iranduba) – and they are precisely those where inequality and the imbalance between humans and the environment is most acute. What we are seeing here is not a relationship of cause and effect but the socioeconomic and structural fragility of healthcare that accompanies the high mortality rate in these communities. The galloping increase in cases of tuberculosis and COVID-19 infections is being reported just where the backhoes of the illegal miners contribute to the mercury poisoning of thousands of people in the Amazon. In a savage irony, these same machines today are excavating shallow communal graves for the victims of COVID-19 in Manaus.

No quarantine was declared for those employed in illegal deforestation and mining. They obeyed no order to take time off work. Contrary to all environmental prescriptions, this destructive pathology has intensified at a time when Indigenous communities are becoming ill and locking themselves down inside their territories, afraid, and with no way of resisting the invaders. Around and even within Indigenous areas, illegal mining proliferates, driven by the brute force and violence of men who risk their lives for dreams of wealth. They are not deterred by any virus. The lowly miners (men who are themselves victims of the absence of public policy for land use and employment) are also a fragile link in the parasitic chain that exposes them to death, even as it feeds the wealth accumulation of their masters.

Although COVID-19 affects people of all races and social classes, without distinction, not all are equally harmed. According to Cameroonian historian Achille Mbembe, writing in 2019, we are living in a time characterized by an 'unequal redistribution of vulnerability.'[113] Indigenous people present greater vulnerability to the disease for both biological and sociocultural reasons. Juxtaposed with this is the precarious situation of the healthcare system. The worst ratios of ICU beds and doctors to inhabitants are found in the states of the Amazon, where both people and the forest are being stripped away. For example, it takes an average of 15 days for medicines and other medical equipment to be delivered to a city in the interior of the Amazon – if delivered at all. The Amazon has become 'the end of the world' when it comes to providing healthcare to its inhabitants, but 'the beginning of the world' in terms of the provision of a cornucopia of forest riches and easy profits.

COVID-19 has further exposed the raw wounds of certain Amazonian populations and of an exhausted forest. Now they are the targets for punishing and possibly fatal blows in the form of normative instructions, decrees, plans, and a GLO. The legacy of these measures remains unknown. What can be stated is that this succession of events is triggering a chain reaction: environmental devastation leads to the emergence of diseases and large reductions in populations, while the environment itself becomes susceptible to an incurable pandemic of destruction.

There is an old Amazonian proverb that says, 'Deus é grande, mas a floresta é maior' ('*God is great, but the forest is bigger*'). It remains to be seen if anything will be left of the forest. COVID-19 seems to have found in Brazil the most favorable comorbidity for its lethal fury: the Brazilian state.

X. Healthcare Means Going to the Community[114]

This chapter gives some concrete examples of the best and worst in Indigenous healthcare. It was written after the end of the Bolsonaro presidency and can therefore include some of the ideas and policies of the new administration of Lula. It draws on a theoretical article[115] written in 2020, on the need for alternative models of healthcare in the Amazon.

In January 2023, the Brazilian Minister of Health, faced with a region-wide scenario of malnutrition and absence of medical care, declared a state of emergency in the Yanomami Indigenous Territory. The Yanomami are the largest, relatively isolated Indigenous people in Brazil and together with the Yanomami in Venezuela their land comprises the largest forested Indigenous territory in the world, measuring approximately 17.8 million hectares. Their territory in Brazil – located between the northern fringes of the rainforest and the mountainous border area – is home to over 30,000 Indigenous people, while another 15,000 live across the frontier in Venezuela. Recently, around 20,000 illegal gold miners and prospectors have invaded the Yanomami's habitat in the rush to commoditize the Amazon rainforest. The results have been immediate and catastrophic: rivers have been polluted with mercury and precious forests have been levelled by loggers and miners. The Yanomamis' ancestral way of life is now endangered and their very existence as a people is under threat.

The Yanomami experience

When he took over the presidency of Brazil from Jair Bolsonaro on 1 January 2023, Luiz Inácio Lula da Silva ('Lula') declared that he was horrified by the situation in the Yanomami Home for Indigenous Health (CASAI) in Boa Vista, (capital of the Brazilian state of Roraima). This is the shelter where Indigenous peoples of the Yanomami, Sanoma and Ye'kuana ethnicities stay when forced to travel to the state capital in search of medical treatment. Lula stated: 'Healthcare should go to the community, and not wait for them [the Indigenous] to come here to the city.'

Lula is right: caring for Indigenous people in their own territory is beneficial in various ways. However, it is not enough simply to send healthcare teams into the community. They must first be provided with the appropriate infrastructure and training, and equipped with a humanist model of care that recognizes and respects the local culture. Hence, the requirement for sufficient investment. The good news is that this is perfectly feasible, as demonstrated by the past experience of the Zo'é people, situated in the northwest part of the state of Pará, as we shall see later in this essay.

For the Yanomami, however, there are a number of urgent questions that demand a response. For instance, why have more than 570 children died from hunger and treatable illnesses in the last four years? Why has the explosion in the number of cases of malaria and of malnourishment not been addressed? Why, as the press has revealed, were 21 requests to help these Indigenous people ignored? Was the Special Secretariat of Indigenous Health (SESAI) not looking after the Yanomami people? Were healthcare professionals not visiting the community? Were we not spending millions of *reais* (the Brazilian currency) to transport patients by air to the nearest hospitals?

The answer is that care was provided throughout this time, but under sub-standard conditions, in an environment poisoned by the ill-will of the (previous) administration of Jair Bolsonaro, which had promoted a series of measures to undermine Indigenous rights and enfeeble the agencies established to protect them. During one of his weekly Facebook broadcasts[116], Bolsonaro stated: 'Indigenous people are undoubtedly changing ... They are increasingly becoming human beings just like us'.

Two factors in particular can explain the tragic scenario that has unfolded in the Yanomami Indigenous Territory. The first is the current medical model, which is distinguished by inequality and attuned to the logic of predatory capitalism. Healthcare is treated as a standardized, individualized commodity, focused more on treatment than prevention. Throughout all these years, the subsystem of Indigenous healthcare has not been able to encompass the cultural particularities of each Indigenous people. It has never been, therefore, an adequate model for dealing with Indigenous populations. The second factor is the invasion of Yanomami territory by thousands of illegal miners – aptly named 'the commodity people' by Indigenous leader Davi Kopenawa Yanomami. These invaders contaminate the waters, indirectly causing the death of Yanomami children and helping destroy the entire Yanomami culture and way of life. They have robbed the Indigenous people of their socioeconomic autonomy, rendering them dependent on the rice and cookies and soda sold on the outskirts of their community.

Lula has responded to the crisis by declaring: 'A way we can resolve this is to establish a task force, you know? There in the community, so we can look after them there. It is easier for us to transport 10 doctors than to transport the 200 Indigenous [people] that are here.'

'Culturality' in healthcare

This 'new' yet actually 'old' way of practicing healthcare can be explained by considering the concept of 'culturality'. Modern medicine confronts a 'crisis of paradigm' which forces us to consider its 'culturality' – the way it operates within the professional, social and economic context in which it is practised. The present model is one of domination: mega-structures support particular models of care and treatment; disciplines are highly specialized; resources flow to the places where medical power is concentrated, not to where they are most needed; medical treatments supplant other forms of knowledge and interventions in human health and the human body; excessive 'scientification' erects walls of indifference between doctor and patient, distorts medical science and its purpose; and market competition rules the interactions between medicine and the human collectives that depend on medical assistance[117].

Thus medicine has become *hegemonic* knowledge, legitimized by science, and is institutionally positioned as the answer to the needs of every individual, group and class. It carries an almost-religious influence, blended with prestige, reasoned confidence, and faith, in its life-and-death decisions. It accumulates symbolic assets, reinvents the reproduction of natural life, and intervenes in the social regulation of habits and customs. In short, modern medicine is the agent and product of its own knowledge, of the 'disenchantment of the world'[118], and of truth.

By contrast, the idea of solving as many problems as possible in communities and districts is not new. Soon after the Cuban revolution[119], a global movement began, mobilizing around the concept of healthcare services organized close to people's communities and homes[120]. The Alma-Ata Declaration, a result of the International Conference on Primary Health Care held in Kazakhstan in September 1978, underscored the fact that 80 per cent of health problems could and should be solved where they occur – outside hospitals. However, the model was not modified over time: it was not adjusted to take account of climate change[121], nor did it incorporate new technologies as they appeared.

Culturality is a necessary revolution, especially in the Amazon[122] and specifically in the subsystems of Indigenous healthcare[123]. Throughout our history, medical practice has saved many lives, but has killed many cultures. Thus, we should emphasize that Amazonian cultures are not only of interest in terms of their specific anthropology and sociology, but they also interconnect with social technologies of other societies around the world. Evidence of this is that we are learning the hard way that combating COVID-19 would be much more effective if we had small healthcare centers and even small hospitals to take care of less serious cases or provide time for convalescence. We have centralized activities in large hospitals, all of which have entered a state of near collapse – as has Brazil's medical system as a whole. Indeed, large hospitals have become incubators of disease as patients hospitalized for conditions other than COVID-19 become infected by the virus, as do the healthcare professionals who treat them. In turn, ambulance drivers – the network in place to transfer patients to large hospitals – have been hit by the virus and have also become sources of infection.

Medical and healthcare systems constitute one of the most powerful of human activities, but our actions need to be balanced and in harmony with the biology, psychology, and especially the culture of the patients and their communities. We should reduce the side effects of the biomedical model of healthcare on individuals, the climate, and the environment, which in turn causes more diseases. Only then will modern healthcare truly represent a source of healing and reduced suffering for Indigenous communities.

Going to the community

The main tool of a 'new' model should not be big city hospitals but small community health centers. Unfortunately, what we have seen in Brazil, in most cases, is just the opposite. We are equipping hospitals, not communities. We are producing specialists who know how to do heart transplants, but we lack those who see culture and the environment as part of treatment.

It appears that Lula and Minister of Indigenous Peoples Sônia Guajajara understand that 'going there to treat' the Yanomami presupposes establishing a minimal structure in their territory to provide medical care and minor hospitalizations. For example, medical technology should also be integrated into the community, and front-line professionals should know or learn the Yanomami language and culture, establishing healthcare teams with the willingness and the time to coordinate the practice of modern medicine with that of Indigenous medicine. Restricted-use medicines, such as antibiotics and even venom antiserums that can be used by the local medical team, should be made available, as well as basic 'over-the-counter' medicines.

Going to the community means investing in satellite communication and in small solar energy generators. The medical teams should also be fairly consistent in their makeup to gain the confidence of the people who are being cared for. Today, medicine enables us to perform an ultrasound using a small transducer connected to a cellular phone. Surgical drapes have become light and disposable. Laboratories are compact, the size of a shoe box. All this technology can and should be placed at the service of people like the Yanomami (who, in turn, care for the health of our planet). At the same time, medical activities need to be humanized to bring them closer to people.

There is one particular precedent[124]: the NGO Urihi Saúde Yanomami set up an excellent health care system in the Yanomami territory in 1999 based on bringing health care teams into the communities, training Yanomami health workers and working alongside Yanomami shamans. They reduced deaths from malaria to zero, and hugely reduced infant mortality. This is another part of the scandal and tragedy of the Yanomami – there was a good health care system operating there first under the Comissão Pró Yanomami (CCPY) then Urihi which the government closed down in 2004.

The Zo'é experience

As well as the Urihi system described above, such a culturally based healthcare model has existed in practice for more than 20 years in the Indigenous territory of the Zo'é people in northwest Pará. The Zo'é model of healthcare is designed to resolve as many problems as possible in the forest itself, avoiding contact with the city and, thus, with epidemics and the prejudices of non-Indigenous people. This practical experiment has been financed since 2000 by the Ministry of Health in partnership with the National Indigenous Foundation (FUNAI). On various occasions, the SESAI team has assisted FUNAI in resolving conflicts among the Zo'é. At other times, it is the FUNAI team that helps healthcare agents make the best therapeutic decisions under given circumstances. The Zo'é are always informed about what is going on, and one way or another, participate in the decisions.

The care given to the Zo'é people is guided by three fundamental principles. The first is respect for the culture and the specific characteristics of the group, taking into consideration the taboos and the traditional medical knowledge of the Indigenous people. The second is to minimize morbidity and mortality risks. As such, removing patients from where they live is avoided, as they have low immunity to diseases from outside. The final principle is the need for institutional covenants to construct this policy, enabling volunteers to professionalize and work as part of a multidisciplinary team. This model also seeks to expand Indigenous knowledge regarding non-Indigenous diseases and medical practices, in addition to promoting a dialogue between traditional knowledge and Western knowledge. During the process, the Indigenous learn about the relationship between epidemics, environmental health, and contact with people from the outside.

The healthcare teams film and make audio recordings of their interactions with the people in the Zo'é Indigenous lands, including conducting interviews about traditional methodologies and local techniques and medications. They also record Indigenous healing practices, as well as the populations' health prior to contact with doctors on the healthcare teams. The recordings capture how the Zo'é perceive 'white man's medicine'. The aim is for this material, which has been collected since 2016, to be analyzed and then systematized as a database that can be placed in the service of the Zo'é people and their healthcare needs. In the future, it may help in developing an educational healthcare program.

It is difficult to compare the situation of the Zo'é with that of the Yanomami. The Zo'é number only around 300 and their territory is much smaller. They are peoples with different histories, who do not have the same level of contact with the rest of society. One thing is certain however: the territories of both peoples are the target of numerous economic pressures. The lust for gold also impacts the Zo'é. So, why are the Zo'é not experiencing a crisis of malnutrition? Why has their territory not been taken over by illegal mining operations?

One reason is that when it comes to protecting the Zo'é people, the work of FUNAI staff has been thorough and highly efficient, and healthcare has always been a part of this work. When there was an outbreak of malaria on Zo'é Indigenous lands in 2006, FUNAI's medical reports helped the agency obtain a state decree that created a buffer zone around the Indigenous territory in which certain activities were restricted. This zone proved to be essential in containing mining activities and, more recently, in protecting the Zo'é from the pandemic. Back in 2016, for example, a small mining operation that had started up close to Zo'é territory was rapidly dismantled by the Federal Prosecutor's Office, leaving no opportunity for miners to prosper. In Yanomami territory, on the other hand, numerous reports of invasions have failed to result in concrete actions on the part of the Brazilian state and, as the years have passed, the government has lost control of the territory.

In contrast, the interaction between SESAI and FUNAI has enabled the conviction of at least one white man who had used Zo'é Indigenous as slave labor in his local nut production enterprise. Later, health reports, supported by information supplied by FUNAI led to the closure of the nut groves during the pandemic, thus limiting the movement of people in the region. The consequences of the policy are undeniable: to date, COVID has not crossed into Zo'é indigenous lands (*Covid não entrou na Terra Indígena Zo'é*). At the time of writing, only three Zo'é have caught the disease, and this was due only to the fact they had to leave their territory in December 2022 to travel to a regional hospital to be treated for very serious illnesses.

The Zo'é experience has worked and should be followed. This care model uses Indigenous territory and culture as barriers against epidemics. If there had not been a small hospital there with basic equipment, the Zo'é might not have been able to resist a Bolsonaro government, whose leader appeared to be working actively on the side of the virus.

To conclude, we need to pluralize our ways of thinking about Indigenous healthcare. 'Going to the community,' as Lula suggested, means much more than taking a healthcare team inside the territory; it involves community integration on many levels. If correctly implemented, it could represent a new way of caring for and promoting Indigenous health in Brazil. It is important for us to support and encourage a healthcare system that sees the forest itself as the largest and most well-equipped hospital a native people could possess.

XI. Brazil's Yanomami People: Silence, Devastation, and Fear[125]

This 2020 essay looks at the current situation of the Yanomami people, made even more precarious by the rapid transmission of COVID-19 among an already vulnerable population due to illegal incursions given the green light by the Bolsonaro government.

Mining for gold and souls

There was a great moment of silence on the banks of the Uraricoera River when Macunaíma, the 'hero without a character,' was born in the depths of the virgin forest, 'black as calcined ivory' and 'sired by the Terror of the Night.' It was there, according to Mário de Andrade, author of the famous 1937 Brazilian modernist novel *Macunaíma*[126], that the Tapanhuma Indian gave birth to the hero of his story.

However, in 2020, the silence of the Uraricoera River Andrade memorializes in his novel has given way to the roar of the machinery of illegal gold mining along its banks. Aerial photographs (taken 18 April 2020) reveal activity by illegal gold miners in various parts of the Yanomami Indigenous Territory in the state of Roraima. The region covers approximately 9.6 million hectares in Brazil and is inhabited by over 30,000 Yanomami and Ye'kwana Indigenous peoples, spread across 360 communities.

Uraricoera River - Photo: Marcos Colón

Rich in gold deposits, the Yanomami Indigenous Territory attracts not only merchants of gold, but merchants of God. The area represents the largest forested Indigenous territory in the world, and it has recorded a steady increase in both illegal gold mining and religious missions. The perpetrators of these activities have been encouraged by the discourse of members of the Brazilian government, which has facilitated the activities of both groups on Indigenous territories by, on the one hand, appointing Ricardo Dias, former member of the Brazilian New Tribes Mission (MNTB), as coordinator of the unit for uncontacted tribes in the National Indigenous Foundation (FUNAI), and, on the other, by its attempts to legalize gold-mining on Indigenous territories. Thus far, the government has been blocked by Congress; however, Draft Law No.191 envisages regulation of mining, exploration of hydrocarbons, and installation of hydroelectric power stations on rivers on Indigenous lands.

The current incursions of miners and missionaries are nothing new for the Yanomami and Ye'kwana peoples. In the 1980s, their territory was invaded by approximately 40,000 gold miners, who left a permanent trail of environmental and human destruction in their wake. In the early 1990s, a group of illegal gold miners murdered 16 Yanomami, including children, in the community of Haximu. The presence of evangelical missionaries in the region of Mucajaí dates back to the 1960s, following the example of Italian and Salesian Catholic missions to the Yanomami in Catrimani, Roraima, near Pico da Neblina. It appears that the Yanomami Indigenous Territory has become the present-day El Dorado for opportunists seeking the 'prizes' of gold and souls in a post-coronavirus world.

The removal of government protection

In early April, several Yanomami reported an increase in illegal gold mining as a result of the reduced presence of state field agents because of the COVID-19 crisis. In fact, the dismantling of assistance in the region had already begun with 2015–2016 closure of FUNAI's three Ethno-Environmental Protection Bases (BAPEs) for isolated and recently contacted Indigenous peoples in the Yanomami Indigenous Territory. The closure made way for the July 2018 murders of two isolated Moxihatëtëa Yanomami by illegal gold miners on Indigenous land in the region of Serra da Estrutura.

The Yanomami and Ye'kwana peoples have consistently complained about the presence of illegal gold miners on their land, says Luis Ventura,[127] a member of the Indigenous Missionary Council (CIMI) in the state of

Roraima. He adds that the state is aware of this and believes the current situation, given the unfolding COVID-19 pandemic, is all the more concerning. The federal government is responsible for implementing territorial monitoring and protection measures to reopen, organize, and operate the protection bases, remove trespassers, and protect the life and health of Indigenous peoples. However, FUNAI has failed to comply with a court order requiring it to reinstate the BAPEs in the Yanomami Indigenous Territory. Its attitude has heightened the level of insecurity in the region, especially now, during the pandemic. FUNAI's failure to implement territorial monitoring and protection measures might well have allowed for the murder of the Moxihatëtëa Yanomami.

In August 2019, FUNAI established a timeline anticipating the reopening of the BAPEs in three stages, one for each base, beginning in August 2019 and culminating in 2021. The project encompassed the purchase of materials, the fight against illegal gold mining, the support of security forces, and the construction of each BAPE. The first, BAPE Demarcação, responsible for the demarcation of Indigenous lands, began operations while its office building was still unfinished. By November 2019, FUNAI had taken measures against illegal gold mining, with the support of the army, including the installation of barriers on the Mucajaí River. According to the Federal Prosecutor's Office (MPF), in March 2020, before the pandemic, the construction of the second base, BAPE Korekorema, was dependent on the conclusion of bidding processes. FUNAI claimed it did not have enough funding to implement the third base, BAPE Serra da Estrutura. When consulted, the regional FUNAI office in Roraima could not confirm whether the Korekorema base would re-open. However, the MPF also said the process of re-establishing the bases should not be the only measure taken to combat land invasion. It asserted that the federal government should have a contingency plan to prevent illicit gold mining.

The Federal Prosecutor's Office in Roraima (MPF-RR) had already warned of the threat of genocide of the Moxihatëtëa peoples in 2017, and filed a class action lawsuit requesting a preliminary injunction against the federal government, FUNAI, and the state of Roraima, so that:

> *[...] the necessary measures be taken in order to quickly reestablish*
> *permanent activities in the Ethno-Environmental Protection Bases ...*
> *providing the material and human resources needed to monitor and inhibit*
> *the operation of illegal gold miners in communities and to guarantee the*
> *wellbeing of the local population and the preservation of natural resources*
> *on Indigenous lands.*[128]

Virus-related unemployment fuels illegal activity

The audacity of the current incursions – in the middle of a pandemic – is shocking. However, what is of even greater concern is the imminent threat the increase in the number of illegal gold miners poses to the health of Indigenous peoples. Social distancing measures and economic restrictions imposed as a result of the pandemic has led to high levels of unemployment. Hence, many more people have resorted to working illegally in this sector, and it is believed that this illegal activity in Indigenous territories will intensify further as a consequence of the economic impact of the virus in the cities.

The presence of illegal gold miners in the Yanomami Indigenous Territory could lead to transmission of the virus, jeopardizing the survival of an Indigenous people that not only has a history of low immunity to western diseases, but also lacks access to adequate healthcare. As Dário, son of Yanomami leader Davi Kopenawa and vice president of the Hutukara Yanomami Association (HAY), says:

> We are very concerned about the disease reaching our communities. I hope it doesn't come to this, but it is very close, in Roraima. We have very serious problems with trespassers, illegal gold miners, coming into the Yanomami Indigenous Territory on a daily basis. They will transmit the disease to many of us. They constitute the main transmission factor. We are afraid about what might happen today, tomorrow, or afterwards. We are extremely vulnerable.[129]

'There is no longer anyone left'

The narrative of Macunaíma draws attention to the violence experienced by the mythical/narrative space which is the Amazon Forest. In the end, what makes the protagonist Macunaíma leave his people is the theft of the sacred Muiraquitã amulet by a character who symbolizes exploitation of the forest: the merchant, Venceslau Pietro Pietra, who goes to São Paulo to enjoy the good fortune the Indigenous talisman has bestowed on him. However, after many twists and turns, Macunaíma recovers the sacred artefact and returns it to the heart of the jungle. In addition to the narrative's fantastical elements, Macunaíma's experience when he returns is of a place he finds unrecognizable. It has been devastated by predatory practices and penetrated by other cultures, which transform the hero while destroying his home and native peoples:

*There is no longer anyone left. Sorcery and bad luck have finished off the
scions of the Tapanhuma Tribe, one by one. The places they knew — those
spacious savannas, those clefts and gullies, those balata bleeders' trails, those
abrupt ravines, those mysterious forests — they are all now as solitary as a
desert. An immense silence slumbers over the Uraricoera River.[130]*

In the fictional space, this destruction takes the tragic form of Indigenous
genocide in the narrowest sense and, more broadly, the annihilation of
nature, consisting of interactions between humans and non-humans. Only
silence, devastation, and fear remain.

The ecological problem, therefore, manifests not as a fact on the textual
surface of a physical tragedy, but by means of the story's conclusion,
which asserts there is no space for either Macunaíma's people or his
own existence as a culturally diverse being. His ascent into the heavens is
symbolic of his displacement in the world. Thus, there is no longer a hero
to defend the forest, no longer a people to be defended. What is left is
just the story, the tragedy, the legend, as fragile as the current condition of
the Amazon rainforest. Our concern is that the same will happen to the
Yanomami and Ye'kwana peoples.

The Yanomami land is experiencing a severe humanitarian crisis.
According to Survival International[131], miners continue to operate illegally,
with incidents such as the imprisonment of three young Yanomami boys
by miners. Dario Kopenawa Yanomami, vice president of the Hutukara
Yanomami Association, reports that despite an emergency decree, the
situation remains unresolved and children are dying from diseases like
malaria, pneumonia, and tuberculosis.

Today, mining controlled by organized crime is rampant with criminal
factions such as the PCC (First Capital Command) and the Comando
Vermelho (Red Command) directly involved in mining operations or
controlling the opportunist groups which flock to the area in search of
rapid enrichment[132].

The violent resistance mounted by illegal miners with ties to these
criminal organizations presents a formidable obstacle to effective removal
efforts. Additionally, recent incidents of violence, including attacks on
Yanomami villages and the murder of indigenous individuals, underscore
the urgent need for sustained action to protect the Yanomami people and
their land[133].

Notwithstanding the measures taken by Lula's government, the
situation of the Yanomami remains very serious. Fiona Watson from
Survival International highlights that, despite the President's promises, the

operation to remove miners has been ineffective. Miners are returning, old sites are being reopened, and essential health services are not functioning. The situation for Yanomami people across the border in Venezuela is also dire, with miners supported by the Venezuelan military and a severe malaria epidemic claiming many lives[134].

XII. Above the Marombas: The Pandemic in the Amphibious Amazon[135]

This short piece, written in 2021, succinctly shows how the historically high level of flooding that year (the forest's 'response' to exploitation) exacerbated the pandemic's devastation of the region.

In June, during the second year of the new biological or biocultural plague ravaging humanity, I crossed the wooden bridges of the small Amazonian community of Cacau Pirêra, in the municipality of Iranduba, located in Brazil's northern state of Amazonas. With a strapless camera in hand, fearing I would fall with each step, I negotiated the *marombas*, in order to spend the day getting to know this resilient floating community, who live just a few miles from the capital of Manaus.

The *marombas* are wooden walkways running above water level that have been erected by the inhabitants to connect their houses. They are typical of the Amazonian region. My nervous attempt to navigate these 'roads of wood,' which are ramshackle but extremely effective at keeping one dry and free from contact with the stagnant water below, formed a counterpoint to my meeting with the residents, whose calm smiles seemed to say, in the face of the floods and the pandemic, 'It will get better; we've seen worse.'

For the Cacau Pirêra community, living along the banks of the river and experiencing all its idiosyncrasies is an ambiguous and often precarious matter, one that emphasizes their sense of exclusion. Indeed, according to the Amazonas Civil Defense Unit, 58 of the state's 62 municipalities have been hit by floods in 2021, and around 450,000 people have been affected. This is equivalent to one out of every ten people in the country.

Then there is the terrible impact of the virus in Amazonas. This disaster, however, is even worse when we look at Brazil at a national level, where the number of deaths at the time of writing has surpassed 534,000. A large proportion of these deaths could have been avoided if Brazil had followed – to an even remote degree – the international examples of vaccine acquisition and use. What some would consider genocide has been heightened by the disastrous policies of the Jair Bolsonaro administration, which continue to underestimate the true number of victims, particularly as so many deaths remain unaccounted for.

Added to this, in the last few days, the local papers in Amazonas (there is little to no media interest in the region at a national level) have traced the dramatic rise in river water-levels in the northern state. This is, historically, a natural phenomenon, but this year, on the first day of June, the Rio Negro reached a high-water mark of 99 feet in Manaus, the highest since measurements began almost 120 years ago. Downtown and suburbs of this so-called 'Paris of the Tropics,' where business is concentrated and people throng the wide avenues and narrow lanes, floods have already caused some shops to close. These establishments had already been rendered vulnerable by the pandemic.

Leaving aside the issue of climate change, these extreme floods have increased problems for the already fragile Amazonian public health system – a microcosm of the situation affecting the entire Brazilian healthcare system. Floods can also cause massive social and economic impacts including loss of infrastructure, destruction of property and important sectors of the local economy, such as agriculture and harvests on the river floodplains, and devastation of the livelihoods of the people who live there. It could be argued that floods are the forest's response to the violence wreaked upon it.

An escalation of the COVID-19 pandemic would worsen these problems. Displacement caused by the floods would force people to move in with friends and relatives in houses with little space, in environments overcome by the waters, or to share hospital wards, or find shelter in gymnasiums with neighbors who are

also affected by the floods. This type of situation ends up magnifying public health risks, facilitating infection, and fostering the potential for the spread of new variants of the coronavirus, as national and international entities have been warning for some time.

However, there are those who resist adversity with conspicuous determination. This is the case of the owner of a small kiosk selling fruit at the Manaus Moderna Market. The trader, Edvaldo Diniz, emphasizes the practical collective experience of men and women used to living with the river's depredations: 'We took up a collection with our co-workers to buy wood and build flood barriers to find a way to keep selling. Of course, we are afraid of an accident, but we have no choice. The level of the river is frightening, it is rising really fast.'

On the other side of the Rio Negro, which surrounds Manaus, I arrive in the town of Careiro da Várzea. Its problems are the same, or possibly worse as regards infrastructure. The town is part of the metropolitan area of the capital, and the families living there have been punished by the floods. Many are homeless and have lost their essential means of subsistence, such as agriculture, one of the main sources of income in the region. One of the river-dwellers of the Juma community, Abrahim Seabra, told us how the rising rivers have resulted in the partial loss of their crops:

> *Almost every year, in big floods like this, we lose crops: cassava and other plants like cupuaçu. We lost everything this time due to the force of the waters and the rising river. The water keeps rising. We are going to have to take the vegetable patch out completely because we'll go hungry if we dont sell anything.*

Knowledge about the nature of floods and tides is part of life in the Amazon, and from this knowledge emerges the poetic nature that sustains the region's imagination. The remarkable sense of equilibrium I experienced in just one day is part of the daily life for the people of the *marombas*. Only if we keep our eyes are open to their wants and needs will we be able to understand these resilient Amazonian amphibians.

XIII. The Amazon and the Enigma of 'Pure Luck'[136]

The following piece, written in 2020, sees the tragedy of the pandemic in the Amazon as part of a constellation of politically motivated disasters in the rainforest.

Quando o céu estiver preto
E das nuvens até as sombras assombram
Quando o céu estiver preto
E das nuvens até as sombras assombram
É só o reflexo do que está acontecendo
Só está faltando fósforo. Me dê aí!
Não esqueça que nesse momento
O vento sacode as árvores
E o clima que fica e o ar agitado
Dizendo tudo o que pode acontecer

When the sky is black
And even the shadows of the clouds are haunting
When the sky is black
And even the shadows of the clouds are haunting
It is only the reflection of what is happening
Only a match is missing.
Give it to me!
Don't forget that right now
The wind shakes the trees
And the atmosphere that remains and the agitated air
Saying everything that can happen.

'Os Pingo da Chuva' ('The Raindrops') is a song by the Novos Baianos, released in 1997, which reappeared on social media when another Brazilian musician, Nando Reis,[137] released it on video in May 2020. In the recording, he rehearses the song, an old favorite, with members of the original group.

However, other things connect the 1997 song lyrics to 2020. One of them is the fact that the sky may be about to go black once again, as happened the previous year[138] during fire season in the Amazon, with its exponential peak in fires[139]. We may soon be reliving similar moments.

Everything is in place for a repeat of this scenario. In early June, the National Institute for Space Research (INPE) reported data it had received from the Satellite Monitoring of the Brazilian Amazon Forest (PRODES), a program that publishes official deforestation data on the Amazon. Data showed that 10,000 km^2 of forest had been destroyed, a 34.2 per cent increase compared with the same period in 2019. Furthermore, in the midst of this record deforestation, fines for incursions and illegal logging fell by 50 per cent. Indeed, the government has been channelling every possible effort to 'let the stampede through'[140] and help land invaders, while attempting to discredit INPE's data. Last year, Ricardo Galvão, then-president of INPE, was finally exonerated after President Bolsonaro had accused him of being 'at the service of NGOs'.

Of course, the press has had to prioritize coverage of the rise and spread of COVID-19. But it is important to contextualize and suggest possible ways of analysing the pandemic based on scientific and political debates. This, the biggest health crisis in recent history, is part of an episodic violence, which revolves around other fundamental problems facing the Amazon region, including the greatest of all – government indifference. For example, poorly structured, weakened healthcare systems could suffer even more pressure because of the fires, leading to further deterioration of people's respiratory health, already compromised by the virus. With the arrival of the dry season, those responsible for deforestation return to the deforested regions to 'finish the job' through further burning.

In less than two years as president, Bolsonaro has managed to confect an image of Brazil in the international sphere which is both troubling and unparalleled. With a man in the role of Minister of the Environment who has overseen a succession of attacks on environmental legislation, Bolsonaro has nevertheless complained on Twitter that the country has been unfairly pilloried for its environmental record.

Yet not even the pandemic has been cited as a reason to interrupt the planned devastation. In the magazine *Época*[141], the former director of the Brazilian Forestry Service calculated that the new coronavirus should realistically be a vector for diminishing deforestation, but the government's dismantling of the environmental surveillance agencies (now full of military police and other such figures), the implementation of official land-grabbing projects, and the encouragement of illegal incursions into the forest have instead helped maintain the scourge of deforestation. The same article contained even worse news: the fire season would coincide with the peak of the pandemic in the country. Nevertheless, a report from *O Eco*[142] indicates that the Brazilian Institute of the Environment and Renewable Natural Resources (IBAMA) delayed by two months publication of a public notice for the hiring of firefighters for the Prev-Fogo (Prevent Fire) program.

In relation to the illegal seizure of land in the Amazon, it is no surprise that Juliano Baiocchi Villa-Verde de Carvalho, a deputy procurator general in the Federal Public Prosecutor's Office[143] (MPF) who defended the controversial 'land grabbing bill', has been appointed by Brazil's Attorney General, Augusto Aras (a known supporter of Bolsonaro's agenda), to coordinate the 4th Chamber of Environment and Cultural Heritage of the MPF[144]. This is the very office that should oversee and control such activities. Juliano Baiocchi Villa-Verde de Carvalho stated that the bill was an exercise in 'free private initiative.'[145]

Exploration and reparation

Citing economic rationale may not be the most humane way of critiquing the problems of the Amazon, but it is worth noting that, according to the newspaper *O Estado de S.Paulo*[146], 30 international investors have reiterated their concern about environmental policy. The head of the National Council of the Amazon, Vice President Hamilton Mourão, commented in reply that the country will respond with 'truth and hard work.' As Ricardo Abramovay states in his report (later published as a book), *A Amazônia precisa de uma economia do conhecimento da Natureza (The Amazon needs a an economy based on knowledge of Nature)*[147], the Amazon provides a greater return from investing in the standing forest than in investing in successive editions of the Safra Plan (dedicated to agribusiness financing). In the opinion of Ricardo Galvão of *Deutsche Welle*[148], the best solution for now is not legal action but pragmatic business actions, such as those cited by Abramovay, sustainably exploiting the forest's biodiversity by providing product certification in areas of sustainable management and awarding tax incentives to encourage good practices.

Nevertheless, these suggestions beg several questions. How can we think of other policies and sustainable practices for the Amazon in the middle of a pandemic? How can we overcome the disgrace of entire Indigenous peoples losing their tribal elders and their history to COVID-19? What do we have to offer as reparation for the daily losses of these peoples? Is there still time to avoid a genocide of these native populations? How can we go on as a country when we reduce environmental safeguards in exchange for benefits to invaders, and jettison indigenist policy in exchange for donations of basic foodstuffs?

We have to remember that a large part of virus transmission to the Indigenous peoples of the Amazon is caused by these invaders – as is deforestation and the devastating fires. While the coronavirus spreads through the lives of people and territories that should be protected or isolated, these questions need answers. And as INPE scientist Carlos Nobre stated in a recent conversation[149], the relationship between the pandemic and deforestation is closer than imagined:

> *It is still a scientific mystery as to why the severe disturbances in the Amazon have never before generated a major epidemic or pandemic. There are millions of microorganisms and every type of disturbance and opportunities for potential pathogens to migrate to humans. In the absence of a solid scientific explanation, a (non-scientific) way of saying it is: 'pure luck'.*

Part 3

Beyond War: Life in the Amazon

XIV. A Paradise Under Suspicion[150]

This chapter is based on an essay, published in 2020, which argues the Amazon's historical characterization as a 'suspect paradise' conceals the twin barbarities of expropriation and neglect.

There are those who prefer to succumb to the fascination of the Amazon as myth and metaphor rather than consider the realities that nestle within its territory. The fictional and scientific 'invention' of the Amazon – be it through pre-colonial records, Indian mythology, Greco-Roman historiography, or, as Neide Gondim[151] describes, the clash of the medieval mentality with the 'Century of Lights' – denotes an almost generic tendency to embrace the Amazon as a permanent locus of mystery and reverie, the chimeric and the oneiric. According to this vision, the tension between man and nature (read 'nature and culture') is never resolved, and the region continually recreates and relives a never-ending story of submission to the domination of natural forces.

However, now more than ever, the paradoxical character bestowed on the Amazon over the centuries requires examination. This essay's title alludes to the dubiousness of the region's association with the idea of paradise. For some, such as writer Alberto Rangel[152], who referred to the Amazonian Forest as a 'green hell', it is the very antithesis of paradise. Indeed, the exceptional nature of its geography has provoked a counter-imaginary that offers instead a portrayal of the struggle of 'civilized man' to survive in the humid tropics, where disease and a hostile climate condemn to death all who endeavor to live there. Similar imaginaries characterise much colonial writing about Africa, especially West Africa – 'the dark continent', 'heart of darkness' and 'the white man's grave'.

This 'dubious paradise,' captured so well by Leopoldo Bernucci[153], is a vortex of continuous natural destruction: slowly, tirelessly the place devours itself – not only its plant world, but also its animal kingdom, and even the people who live there. An ecocritical reinterpretation of Bernucci stresses how the hostility of this natural environment is aggravated by the hostility of its human invaders, whose greed either destroys or transforms it into 'agrodollars' through a process that Rob Nixon[154] describes as 'slow violence.' The concepts of a 'green hell' and a 'dubious paradise' can be found across literature and history, science and art, and reveal the close relationship between these apparently independent fields. That is why the social and historical processes are both imaginary and real; they follow the development of attempts, in each literary canon or scientific episteme, to refashion Amazonian reality. The fictional and scientific contexts of these interpretations belong to the same sphere and are driven by the same logic of (re)creation.

Poised on the border between fiction and reality, the Amazon region has thus become a bundle of mythical, redeeming, fantastical, sacred, and profane fabulations. Hence, in the opinion of Marilene Corrêa da Silva[155], the 'voices' of fiction mirror the ruthlessness of scientific discourse.

As part of its 'disenchantment', the Amazon continues to be seen as a metaphor for the mythical and the astonishing. This isn't about ignoring reality – that would underestimate the influence of the real world on our imagination. Instead, it emphasizes the various representations and meanings that different disciplines, as well as scientific and artistic narratives, attribute to the Amazon.

This is why scientific voices, frequently associated with arbitrary reconstructions of the region, can be as unrelenting as the voices of fiction, and can radicalize perceptions of the Amazon even more powerfully than fictional composition. In fiction, artistic imagination enjoys the freedom to dissolve and reconfigure frontiers, subjects, spaces, and the nature-culture relationship in a process of 'creative invention.' It transforms living beings (humans and non-human), the sacred and the profane, into hybrid phenomena, and recreates the relationships between them according to the author's own imaginative impulses, as Mario de Andrade does in *Macunaíma*[156].

According to Ana Pizarro[157], the Amazon, as a physical and cultural space, contains elements that act as symbolic devices for the writer. These elements stimulate the imagination, allowing writers to make connections and draw comparisons with what they encounter. The Amazon becomes a mythical universe that responds to their own shortcomings, expectations, and physical and spiritual needs in a way that is subsequently confirmed in their fictional narratives. And if the literary voices grow more merciless, it is because their narratives become a conduit for the intensity of the conflict between the forest peoples and invaders, a conflict that emerges with increasing visibility in their writing.

For example, in Milton Hatoum's novels *Dois Irmãos* (The Brothers)[158] and *Órfãos do Eldorado* (Orphans of Eldorado)[159], both the characters and the city of Manaus are depicted undergoing significant transformations and facing various challenges. These narratives vividly illustrate the economic, social, and environmental impacts experienced by the local inhabitants, whether these challenges stem from external invaders or internal conflicts.

Nevertheless, it is important to emphasize that both the scientific and literary narratives are not autonomous, but rather forged within a political context that justifies their discursive purposes.

'Here starts Brazil'

During my last journey through Tabatinga, a city situated on the Brazilian border with Colombia and Peru (see Chapter VIII), I saw a striking inscription on a memorial to the city's foundation, which proclaims: 'Here starts Brazil'. But this phrase also encourages an opposite reading, in which the border signifies the end of the 'civilized' world: here starts a remote, uncivilized, and barbarous realm that has been left to its fate. This foundational maxim has given rise to a discourse justifying authoritarian control of the region, resulting in the predatory exploration for and plunder of its resources. It is another facet of the narrative of the Amazon as a 'green hell,' which has led to the imaginative and political exploitation of a region designated as an empty space, a *'terra nullius,'* open to the violent imposition of civilization[160].

As sociologist Octávio Ianni[161] relates, in the beginning, there is only nature. Then comes man. Nature and human culture struggle, and one is imposed on the other: equal, unequal, non-compliant. Soon man emerges as the 'lord of nature'. Once it is appropriated, nature is transformed. Man also is no longer the same. Both have lost their innocence and become a part of History. Now, the 'slow violence' of yesterday is no longer so slow; it has permeated the entire region, leaving the future of both humans and non-humans in doubt.

The indifference to, and abandonment of the Amazon by Brazilian authorities during the COVID-19 pandemic appears to be novel. But as Corrêa da Silva argues in her work *Amazon Country (O Paiz do Amazonas)*[162], the region has always been regarded as the most imperfect feature of the nation-state and merely serves as a currency of exchange. It represents a social formation (in stark contrast with the social formations of the Indigenous Amazon) whose relationship with the larger Brazilian nation is guided by the prevailing order of the colonialist state. The present hardships in the region, exacerbated by COVID-19, clearly show the heavy price the Amazon pays to be part of Brazil: it receives less than it gives both in taxation and in political terms, yet it fails to figure among the state's national priorities. As authors Samuel Benchimol[163] and Arthur Cesar Ferreira Reis[164] show, indifference and neglect are also strategies for clearing the territory. As a result of this indifference, many in the region will die and nature will again be free of human interventions.

The traditional literary vision of the Indigenous peoples of the Amazonian region is a romanticized one, oscillating between images of semi-human innocence and savagery. This is their price for inclusion in the national imagination. The scientific vision has focused on their innocent 'otherness' and the uncertainty of their humanity. Meanwhile, the economic vision has been a more overtly brutal one, relegating them to the condition of animal, slave, or inferior subject. The literary narrative of Amazonian Indigenous innocence and brutality, however, is closely connected to the idea of 'uncivilized barbarians.' That is why so few appear to find it strange that this 'dubious paradise' consumes millions of lives, through slavery, disease, starvation, exhaustion and extermination. It is important to emphasize that, in the Amazon, the political embodiment of doubt, as Bruno Malheiro[165] points out, emerges from the idea of the region as a risk to the nation-state's sovereignty.

If this seemingly aberrant 'difference' is juxtaposed to what is meant by 'nation,' it is but a short step to the normalization of violence and a permanent 'state of exception'. The doubt that invented the idea of Amazonian risk and Amazonian emptiness has historically been manifested in the state's use of force and violence which is promoted as the only possible response to these (human and nonhuman) suspects.

Conjunctions: a 'dubious paradise' and a 'green hell'

Contemporary narratives of the Amazon also reflect the historic dispute between scientific and literary approaches to the Amazon and their attempts to describe or transfigure reality. Both approaches reflect the famous intellectual and political disputes that have taken place over more than three centuries of attempted rationalization. A growing dread of intellectual contamination during this period led to the increasing distinction between religion, science, and art, and their separate institutionalized disciplines. Strong rules for a 'scientific spirit' were established, associated with the independence and republican movements. These disputes have placed western literary and scientific thought sometimes in opposition and at other times in harmony with each other.

These differences were perhaps particularly acute in Latin America, including Brazil, because of the identification of leaders of the independence movements with scientific rationality in opposition to clerically backed royalty, aristocracy and tyranny.

In non-literate societies, oral literature plays an important role, filling the void left by the absence of recorded discourse, including scientific discourse. Initially ignored by scientific discourse, oral narratives have been incorporated without due credit. For indigenous peoples, ancestral stories and myths are fundamental, offering a holistic view of life in contrast to the more linear and compartmentalized approach of Western science.

Today, oral literature serves as a space to preserve and transmit ancestral knowledge, establishing a dialogue with scientific knowledge. This interaction is exemplified in narratives such as those of Johann Baptist von Spix, Carl Friedrich Philipp von Martius, La Condamine, or Father João Daniel, João Barbosa Rodrigues, and Davi Kopenawa, where different perspectives confront or intertwine. The account by the authors of *Brilhos na Floresta*, Noêmia Kazue Ishikawa, Takehide Ikeda, Aldevan Baniwa, and Ana Carla Bruno, illustrates this dispute between scientific and literary knowledge. During an expedition in the Amazon jungle, guided by Aldevan Baniwa, they witnessed the spectacle of bioluminescent mushrooms after extinguishing their lanterns to adjust to the darkness. Noêmia described the experience as unforgettable, while Ikeda questioned why he had never witnessed such a phenomenon. Baniwa explained that sometimes scientists need to 'deilluminate' to truly see.

The Amazon is part of this historical invention of intellectual approaches and narrative canons, since, from the time of the meeting between the Americas and the 'old world,' its physiography, its peoples, and its territories came to signify the 'anti-world.' It represented the 'unknown,' and there were no existing parameters for its description; it lay outside the logic and rationality of fifteenth-century art and science[166]. These circumstances have led to the Amazon's inclusion in both artistic creations and scientific discourses. Despite their different approaches and occasional disagreements, these creations and discourses share a common mental framework, where they coexist and interact.

In turn, the presence of this unusual and unexpected American world of 'almost humans,' and their equally strange, distant territories and cultures, profoundly altered the West's scientific and artistic narratives.

These drew upon each other to describe and produce explanations that adhered, to a greater or lesser degree, to facts. Thus, science and literature assisted each other, reciprocally, in *Invention of the Amazon*.[167]

Even today, scientific reports describing the dynamics of the region often adopt literary qualities, while literary works sometimes include scientific details. Despite ongoing efforts to maintain a clear boundary between these disciplines and resist this 'intellectual cross-pollination', both are ultimately struck by the region's nature and culture. They are equally challenged by the historical and environmental conditions of the Amazon, which present significant issues to the contemporary world.

During the course of colonization and nationalization, the literary and scientific narratives invented the Amazon for the rest of the world. The exercise of their considerable influence attracted further narrative disputes over what is known. Today, in the context of the pandemic, we see the rebirth of earlier metaphors of distancing and estrangement in the descriptions of the region and its interpreters – that is, of the Indigenous peoples who have lived there for over 10,000 years.

The 'green hell' and 'suspect paradise' are at the same time, factual and imaginative syntheses of the tropical, humid, natural world that is being devastated by the voracity of invaders and the trajectory of their exploitation of its peoples, biomes, and ecosystems. This 'paradise lost' has acquired the condition of a cruel reality for today's Amazonians. Lacking citizenship, they are orphans of the Brazilian nation-state. The state that the Indigenous inhabitants of the Amazon encounter is not one that generates opportunities and guarantees rights, but one that generates violence through its monopoly of the forces of 'law and order'. The Amazon's condition as a region inhospitable to outsiders has not impeded its penetration by capitalism: its waterways cut through Indigenous territories and environmental reserves and creates municipalities that depend on the circulation of extractive merchandise and products.

This 'dubious paradise' conceals the twin barbarities of abandonment and expropriation. Under the complicity of silence, millions of human lives have been sacrificed, denuded of their cultures and their historical adaptability to sacrosanct environments, through which they humanized nature. Scientific and literary narratives converge in signalling, through denunciation and reflection, the conflicts and contradictions of predatory interventions and their effects over the centuries on the humans, non-humans, and resources of the Amazon.

The pandemic has compounded the centuries-long impact of the disappearance of hundreds of peoples, cultures, and even civilizations through epidemics, genocidal territorial disputes, slavery, and servitude to the logic of capitalism. Will the exuberance of the forest conceal this latest politically driven, sanitized massacre attempted by the Bolsonaro government? Will solidarity journalism reach the intellectuals whose duty it is to record the region's self-awareness of the crisis in search of alternatives for both the local communities and the planet as a whole? The answers to these questions will determine whether the Amazon's story is one of forgotten tragedy or a call to action for global solidarity and justice.

The history of the Amazon is full of brutal episodes of the conquest of 'pioneering new fronts,' and the problems of yesterday continue to be repeated today. But it is the political exclusion of the Amazonian peoples from their constitutional rights, and their physical and cultural heritage, that awakens the most contradictory responses. These include shameful intellectual justifications of government power which attempt to 'explain' the abandonment of the Amazonian populations to the ravages of a pandemic that the 'civilized world' itself produced.

Once again, both literary records and scientific narratives are at work interpreting the spread of the pandemic in the region. The accumulation of these narratives may help to lay bare the deceptive image of a 'dubious paradise' hiding in the depths of the forest and portray it as yet another tragic episode in the nature-culture duality created by the human lust for riches and power. The Amazon of today, however, cannot continue to be enshrouded in myth, much less metaphorized by fables that reinvent the forest according to the Eurocentric imagination. What is at stake is our very existence – that is, the life of all people, whether individuals or groups, whether urban, rural, or Indigenous. The remaining biomes and ecosystems are threatened by economic predation during and after the pandemic,[168] due to the diminishing amount of merchandise available to the central cities of the Amazon's municipalities and the stripping of opportunities from the region's peoples by government policies in education, health, and labor.

Beyond western scientific and literary narratives on the Amazon, however, we find – with 'slow seeing'[169] – the imaginary narratives and representations belonging to its native peoples. These offer a new repertoire of ideas to help us contemplate the post-pandemic world. This world needs new solutions that will enable all of humanity to finally partake in some share of well-being. After all, capitalism's promise of universal well-being as the automatic dividend of 'civilization', never arrived.

Yet who continues to pay the bill for this civilization, which has from the outset confected the image of the 'Indigenous' out of a radical otherness, and who may now be silenced forever? At stake is the survival of us all. By neglecting other ways of understanding the region, we sacrifice the ancestral contributions of the peoples of the forest and, with them, other 'horizons or perspectives,'[170] different to those so far provided.

The end of the Amazon would be the end of the heart of humanity. The heart, not in the biological, but in the ecological (analogous to the oft-presented image of the Amazon forest as the 'lungs' of the world), and poetic sense. The heart in the sense in which both the poetic and scientific imagination can become forces that guide the dialogue between public policy and the peoples of the forest.

XV. Only a Global Coalition Will Save the Indigenous Peoples of the Amazon

In 2020, as the pandemic was raging in Brazil, the healthcare system of Indigenous Amazonians appeared near collapse. This short essay is adapted from an article[171] that called on international organizations to act urgently.

The disturbingly scarce amount of reliable information from the Amazon indicates that the system of medical treatment for Indigenous communities is close to collapse.[172] A national and international coalition for the health and safety of Indigenous peoples and territories is urgently needed to avoid a genocide.

This coalition should have the central objective of addressing the causes of the manifold vulnerabilities of native peoples. These include violent incursions into Indigenous territories, entailing massive and perhaps enduring disruptions to the delicate mesh of biological, ecological, economic, and cultural relations (human and non-human), which in turn may well result in annihilation of the peoples of the forest and riparian communities. It is also necessary to formulate guidelines for an Indigenous healthcare policy that reclassifies the material conditions of the health districts and their relations with healthcare agencies at the municipal, state and federal levels. All levels of government should be engaged in protecting Indigenous lives, as well as the lives of healthcare personnel treating these populations.

There is an urgent need to rethink Indigenous healthcare practices to consider the social, cultural, and economic contexts of the forest peoples. Such guidelines should direct healthcare policies so that, based on care during the pandemic, health treatment centers are established and supported within the Indigenous territories themselves – something that is not happening at this crucial, vitally important moment in the history of the Amazon.

A catastrophic event of this magnitude – especially in the midst of climate breakdown, and the likelihood of more frequent pandemics during this century and beyond, assuming humanity makes it that far – cannot have national limits. Nor can it be construed narrowly as comprising local, isolated, and therefore irrelevant disputes, somehow disconnected from how we live as consumers of what ought to be legally conceptualized and codified as a shared planetary inheritance that is by no means necessarily limited to humans. This mobilization ought to take place at the local, regional, national and international level. It must originate from the consciences of individuals moved to action by their dawning awareness of the situation of the forest peoples, then move like a wave through international civil society, producing the moral and political authority to put the major governing bodies of the Brazilian state on notice. Everyone should be crying out for justice, rights, and the protection of the lives of Indigenous peoples, both in the Amazon and elsewhere.

Lacking access to proper healthcare, already brutally traumatized by the territorial invasions of criminals who profit from the destruction of the forest and its peoples, several Indigenous ethnicities are once again facing the risk of genocide. This situation requires an escalation of activism from the level of 'local' affairs to one of condemnation and a demand for justice at an international level.

Efforts on the part of the Coordination of Indigenous Organizations of the Brazilian Amazon (COIAB) to protect Indigenous communities and advocate for their rights, as well as the rights of various grassroots Indigenous organizations, such as the Indigenous Missionary Council (CIMI), scientific institutions like the Oswaldo Cruz Foundation, and a number of Brazilian universities, are necessary but not enough. Without robust international calls for protection of the living, as well as justice for those already gone, Indigenous lives will be sacrificed to a cruel fate. International organizations and humanitarian institutions – along with the financial, technical, and logistical support they provide – are absolutely indispensable.

Longstanding efforts by the Brazilian state to silence and marginalize Indigenous voices have been honed by the COVID-19 pandemic. Now more than ever it is necessary not only to demand reparations for the lives lost, but also to attempt to blunt the neoliberal predatory practices suffered by Indigenous peoples, the Amazon forest, and its nonhuman inhabitants. In other words, we are asking for political support for the present and future sustenance and survival of Indigenous social life, as well as the necessary conditions to protect the health and demographic reproduction of the various Indigenous peoples that live with and depend on the forest. It is vitally important to recall the role international human rights organizations played in protecting the Waimiri Atroari people from a massacre promoted by the state, which wanted to move forward with construction of the BR-179 highway and hydroelectric power plant at Balbina[173].

The increasingly flagrant disregard for the health of the Amazonian population exemplifies the spurious relations between the federal government and Brazil's north, the part of the country, proportionally speaking[174], most affected by COVID-19. To make matters worse, there is also the problem of underreporting case numbers in the region as well as the degree of vulnerability of local Indigenous[175] communities in the face of the pandemic. COIAB, which keeps an independent count for the Special Secretariat for Indigenous Health (SESAI), states that, up to 15 June 2020, there have been a total of 249 deaths and 3,662 confirmed cases of COVID-19 recorded among native peoples in the region. SESAI recorded 101 deaths and 3,013 infections for the same period.

The pandemic also affects the daily lives of non-Indigenous individuals, who, in some cases, migrate to riparian communities seeking safety[176] in smaller settlements where there are fewer crowds and less circulation. Moreover, irregular river transport[177] has been indicated as one of the factors in the spread of COVID-19 in the interior. The spread of the contagion ravaging rural communities has led to advance of the illness into the Javari Valley,[178] despite warnings by Indigenous leaders. This is the region with the highest density of uncontacted and isolated communities on the planet, with more than 2,000 km^2 and over 6,000 inhabitants, many of whom live in voluntary isolation. We have a moral duty to protect their health and their choice to live without contact with modern society. The Javari Valley is now at a dangerously high level of risk of contagion.

Through a broad, diverse and inclusive coalition of concerned Brazilian citizens, coupled with the solidarity, creativity, and resources of international organizations and institutions, the policies of death that stalk Indigenous communities can be stopped. But there is no time to waste.

XVI. Amazônia Redux: A Re-evaluation of Urgent Needs

The following 2020 essay[179] is a plea to break the silence shrouding the pillage of the Amazon, and to listen and learn from the people of the forest, for the sake of all humanity.

My meeting with a group of Indigenous leaders during a recent trip to the Amazon still resonates with me to this day. It took place in Atalaia do Norte, the riverside entrance to the Javari Reserve, an Indigenous territory in the Javari River Basin. These leaders wanted to share their concerns about the threats to their people, environment, and culture. Among our wide-ranging discussions about ways to defend their people, the words of Tumi Manque Matís from Javari's Matsés tribe stood out in particular:

> *Our land [the Javari Reserve] is important to us. We represent the variety of ethnicities living in a communal system. Although we belong to a number of different groups – Matsés, Kulina, Mayoruna, Korubo, Kanamari and others – in the Javari Reserve, we are all one. We want to keep it that way. The hope of our leadership is to preserve a future for our children[180].*

Tumi Manque Matís issues an urgent call for us to think about Amazonian territory, socio-diversity, and biodiversity in a new, relational perspective, because the divide between nature and culture no longer makes sense. Nature is culture, culture is nature. In reality, there are many Amazons that reveal a variety of tastes, colors, ideologies, and occupations. Unfortunately, today, none of these are part of any overall government plan, the sort of plan that would respect the self-determination of Indigenous peoples and communities, whose interaction with the academic community could generate vital knowledge for the new, more difficult conditions of life on this planet. The Amazon offers a model that generates peace and alliances between people: outsiders and those who live in the forest. As such, it could teach the stewardship of the forest to our citizens, whose precarious lives are currently based on the consumption of objects whose planned obsolescence is built in, and whose manufacture and disposal create continual environmental contamination.

Throughout historical memory, the Amazon region has undergone various re-imaginings by outside subjects. This began with its very name, a tribute to the Greek myth of the Amazons. Its previous history has been utterly ignored, and since this epistemological break (its re-baptism in the 16th century), the region has experienced more than 500 years of hardship imposed by predatory systems of exploitation. Whether it was the exploitation of animal life or plant life, minerals or waters, such practices have often defined the tragic destiny of Amazonian communities. Now, it will again be redefined under President Bolsonaro, with the conservative, neoliberal agenda of his administration. Another round of disruptive policies is being implemented, completely contrary to living a good life ('Bem Viver') as defined by Alberto Acosta[181]. For Acosta, a good life does not stem from an academic-political proposition; rather, it embodies the opportunity to learn from the diverse realities, experiences, practices, and values present in different habitats – even today, in the midst of capitalist civilization. Once again, however, we are witnessing yet another group of subjects from outside Amazonian territory attempting to determine who should inhabit the land and trying to enforce their experiment.

The most recent western science – shared in magazines, books, lectures, and audiovisual demonstrations – suggests that the best destiny for the Amazon and its peoples, as well as for the climate and the biodiversity of the planet, would be for us to maintain and reproduce the complex relationships with the forest practised by the people who traditionally live there – the Indigenous peoples, Quilombolas, traditional riverside communities, fishermen, and those who support themselves by using its land and water. They were living in the forest long before the domestic and international predators of the global world market arrived. Indeed, US Nobel Prize recipient Elinor Ostrom has evaluated and endorsed such common-use systems.[182]

Shrouded in silence

Let us consider for a moment an Amazon that does not appear in the media or in contemporary scientific imagination. Silence appears to shroud the vulnerability of traditional peoples to personal violence, to the sociocultural and environmental dismantling of their habitat, and to the economically and culturally predatory large-scale infrastructure projects the Brazilian government is pursuing in the region. This silence is a key component of a 'civilizing' plan that fails to engage Indigenous peoples, Quilombolas, traditional riverside communities, small farmers and forest workers, or to comprehend the Amazon's cultural diversity and complex biomes. Recognizing the existence of this silence imposes a new duty on us all to critically seek out, investigate, and collectively share this essential information. What is in play is the survival of us all – those living in the Amazon as well as the rest of humankind. When there is nowhere left for the Indigenous, there will be nowhere left for the rest of humanity. Sister Dorothy Stang, who was murdered in February 2005, understood that 'the death of the forest is our own death' – the phrase that was printed on the T-shirt she was wearing at the time of her death.[183]

The crucial difficulty stems from the fact that colonialism did not end with Brazil's independence. Indeed, since that time, the Amazon has become part of a continuous process of renewed colonization. New eras have led to new forms of colonialism: the forest continues to be an outdoor laboratory for economic interests and ventures that fail to observe or question the impact of the invasion of lands and the submission of peoples to the various generations of conquerors. Indeed, even before the contemporary phase of Brazilian society, the state had no specific plan for the Amazon except that of surrendering it for sale to the highest bidder. The Amazon became a currency to be exchanged for any form of experiment or intervention that claimed to be productive – as if keeping the forest standing and reproducing was not productive enough, something Amazonian people have been doing for millennia, preserving the 'flying rivers' that ensure rainfall throughout South America.

The Brazilian government is still uncertain about how to cash in on the Amazon. Its traditional peoples, together with many scientists, have already found several sustainable solutions, but these do not conform to the demands of agribusiness, and the mining and extractive industries present in the region. The Amazon is currently a territory of enclaves – in Mato Grosso, the Manaus Free Trade Zone, Acre, and Pará. These enclaves are oblivious to the dialogue between traditional peoples and the latest western science, which nurtures the richest knowledge of the region, since Indigenous peoples, among other groups, are also producers of scientific knowledge and hold a critical understanding of how the Amazon region could be managed.

Meanwhile, the radical subjugation of the 'Indigenous,' who have been silenced for as long as can be remembered, means they have continually paid the bill for civilization's malaise. Sadly, Indigenous reserves are treated as the new economic and extractive frontier by a Brazilian government that remains eager to allow their territories to be exploited by national and international corporations, with which it curries favor by restricting the already tenuous pro-Indigenous legislation.

Education for survival

In summary, the biggest problem of the Amazon is this slow, continuous violence, exacerbated by the lack of any form of communication between the government, the powerful economic interests involved, and most of the region's inhabitants. Communication needs to be meaningfully pursued, and this can only be achieved if it is based on a scientific interest in the present and future of humankind. Lack of communication subjugates the Amazon's inhabitants, subverts what could truly be a progressive contemporary national policy in Brazil, and sabotages any autonomous plan that seeks to preserve the culture, biomes, memories, and history of the region. Science, traditional knowledge, and memory are aligned against this fabricated ignorance. Both poetic and scientific imaginations are forces that could guide genuine dialogue between public policy and the peoples of the forest.

The people of the Javari Reserve are fighting not only for the forest but for the education of their people as a tool for their own survival. During my conversation with faculty members at the Federal University of the Amazon at their campus in Benjamin Constant, they informed me that 1,414 students were registered in 2019. Of these, 432 come from various indigenous groups, including those who live in the Javarí reserve[184]: 115 Komanas, 34 Kambeba, 1 Kamanari, 40 Kaixana, 5 Marubo, 2 Mayoruna, 272 Tikuna, and 3 Woitoto. The majors they pursue include business, education, anthropology, languages, biology and chemistry, and agroecology. Thus, the Indigenous people of the Javari have much more to teach us besides how to discover new relational perspectives on culture and nature. Their efforts to travel to nearby cities to seek out a western-style education show their desire to learn in order to equip their own communities. Should we not do the same: learn from Indigenous knowledge and communicate their riches to our peers?

XVII. Stepping Softly on the Earth[185]

This essay introduces the author's 2021 documentary film,
Stepping Softly on the Earth, *which highlights the Indigenous
voices resisting the monocultural farming, deforestation, mining
projects and land invasions that are devastating the Amazon.*

This film places those who have been historically pushed to the
edges of our thought at the center: Indigenous peoples, the
protagonists of a possible future, since, as Ailton Krenak says,
'the future is ancestral,' and we need to learn to 'step softly on the earth.'[186]

We were on the BR-155, a few kilometers away from where it crosses
the Trans-Amazonian highway, when we noticed a meatpacking company
sign announcing our entrance into 'a world of sun and pasture.' The view
on both sides of the road was soon subsumed by enormous farms. The
arid environment wrapped the febrile landscape in a permanent cloak of
heat. A few kilometers ahead, we saw a beautiful Brazil nut tree, exuberantly
elegant, a silent witness, an ancestral presence, forgotten by history in the
middle of a field, calling the cattle to gather in its shade.

The colors of the scene invited us to stop the car in the only place
with a hard shoulder, right next to the entrance to a farm. We took the
drone out of its case and started guiding it towards the tree to capture the
scene. Even before the drone gained altitude, a car, which was about to
enter the farm, stopped as it caught sight of us. The driver stared at us for
a long time. Uncertain as to what to do, we quickly landed the drone. At
that particular moment, we felt threatened: in this part of the Amazon,
where leaders and defenders of human rights and nature are frequently
killed, a car stopping in front of you can mean much more than just an
intimidating look. We did not film the scene, but we felt we needed to tell
this story, and others – of an Amazônia suffering under a death plan that
produces fire, smoke, contamination, and blood.

Extensive livestock production, which supplies the meat packers we
saw near the start of the highway and numerous others spread across the
south and southeast of the Brazilian state of Pará, has steadily penetrated
the various parts of the Amazon Forest, spreading fires and widespread

deforestation. Nevertheless, we are in the Amazon, and the life of the forest still calls the living to a dance against the backdrop of this festival of greed and destruction.

A few kilometers away from the tree, which stood as a solitary, silent witness, we met a group of people who are key to the harvesting of this region's nuts. They come from one of the most severely affected Indigenous territories in Brazil. This now consists of a tract of forest marooned between cattle farms, power lines, railways, and streets, flowering like a solitary rose. This is a story that must be told. It is also the story of the chief of the Akrãntikatêgê people, Kátia Silene, who many have wished to silence, but whose voice resonates with the strength of long-established roots. Every day, Kátia confronts the Vale Mining Company, which, in the first three months of 2021, in the midst of the second wave of the COVID-19 pandemic, posted a profit of $5,546 billion, a 2,220 per cent increase over the same period of the previous year. Profits are frequently followed by an untold number of deaths. However, Kátia is not intimidated. Every word she says broadens our ways of perceiving the world.

The forest falls

It was in the Munduruku lands surrounding Santarém, still within the Brazilian Amazon, that we would again encounter the daily tension facing those who insist on continuing to live their lives in the Amazon. After much discussion as to whether we should travel to witness the latest felling of trees – in an area where threats and intimidation occur every day – we decided that we had to face the danger and go.

We slept for a few hours and then, by way of a back route, arrived at the recently deforested area. The situation was so tense that our local guides, fearing the worst, remained in the car, while we ventured out to record the devastation. The damage was at first hidden by a carpet of morning mist, but this was soon burned away by the sun, revealing the trail of destruction. We were on Indigenous land that had been invaded by soy farmers. The area in Brazil planted with soy grew 166.5 per cent between 1999 and 2018. Soy production, like mining for iron ore and other commodities, has continued throughout Latin America, unaffected by the economic decisions of the majority of governments, and irrespective of their place on the political spectrum. However, the statistics do not come close to capturing the real violence experienced by those who find themselves in the way of this expansion.

Only hours before we arrived, chains stretched between tractors had dragged down an immense area of forest. The smell of fallen trees was everywhere – mixed with that of fear. Finding the forest demolished and surrounded by the silent faces of people who no longer feel safe on their own land, we felt as we had when we met the gaze of the driver who had stopped in front of us by the solitary tree. Yet, out of that climate of tension, another character appeared – this story will be told by a tribal chief who dares to resist the violent threats and the invasions of his lands, Cacique Manoel of the Munduruku people.

The acceleration in the rates of deforestation, in the numbers of forests burned, and in the expansion of monocultures and mining is not confined only to the Brazilian Amazon. Nor is the contamination of rivers with waste from economic activities such as mining limited to a single country. However, the Bolsonaro government of Brazil particularly encourages the pillage of material and energy while eroding all legal guarantees that still ensure a modicum of safety to the peoples and traditional communities, as well as to the land under environmental protection. The war waged against the peoples of the Amazon is the political plan of all those governments that encourage capitalist barbarity in the region, be it in Brazil, Peru, Bolivia, Ecuador, Venezuela, the Guianas, Suriname, or Colombia – the nations whose territories abut the Amazonian biome.

Transcending the maps

We travelled through some of these countries, crisscrossing borders invented by nation states and meeting Indigenous peoples and nationalities not limited by the lines drawn on colonial maps. Amazonian ancestry is connected, and its histories entwine in so many ways that, for the people of these lands, the state is a simply the invention of a grotesque and exclusive plan that has always ignored all other possible ways of being. Perhaps that is why the violent expansion of capitalism in the Amazon is always backed by the state. The greatest and most real connection between so many peoples, cultures, and communities is not the state, but the Amazon River, and it was through the meanderings of this river that we arrived at an Amazonian image-in-common – a third story that needed to be told.

On the Peruvian side, we found a tributary of the Amazon, the Nanay River, and a city erected upon it, Iquitos. The Nanay River is the main source of water for Iquitos, but as in so many other places in the Amazon, mining and oil exploration have contaminated its waters, leaving a toxic legacy for its inhabitants. Here, from the prow of a boat, José Manuyama, '*el hombre del agua*' ('the water man'), a Kukama descendant and member of the Water Defense Committee of Iquitos, told us his story. His knowledge was nourished by the life of the river and made us rethink our own. Indeed, our crude paper sketches only sketched the contours of the reality of the personal stories of those who confront capitalist barbarity in the Amazon each day. You cannot pass through these lands without, at some point, coming across someone who has told stories about the Amazon across the radio waves for 40 years: broadcaster Mara Régia. The strength of her voice guided our narrative.

Having finished filming, from Peru to Colombia to Brazil, from Santarém to Marabá, from Iquitos to São Paulo, we then began editing and, for a long time, were puzzled. There was a hiatus, something missing. Of some things we were certain: that we were standing by the people who confront the system of death and destruction that is Amazonian capitalism, and that we were determined to convey the power of their voices. But then we started to change the terms and depth of our engagement. We needed to listen more than speak. Feel more than define.

To delay the end of the world

In reality, we began to understand we were standing face-to-face with Indigenous stories that could delay the end of the world. Perhaps, to express this profound realization, we should turn to the thoughts of Ailton Krenak. His voice sounds sweet to ears full of uncertainties. Like an ancestral echo, Ailton's way of thinking defined an important step in the film: we speak of three different Amazonians, three stories of struggle, but our characters are not defined just by the terms of those who wish to invade their lands – they are so much more than that. Each, in their own way, offers alternative horizons, a way out of the systemic chaos in which we live. Their ancestral knowledge points to a different way of seeing the world, a particular way of feeling and thinking with the Earth. Through their own voices and the territory they embody, each character invites viewers to completely upend their perspective: to see life where it generally is not seen; to hear voices where they are generally not heard; to feel the sensation of 'stepping softly on the earth', an experience that the concreteness of the cities and the hours on the clocks have taken from our lives.

Different paths cross, distinct worlds draw closer together and form a story that needs to reach us. The Indigenous Ku-kama, José Manuyama, lives in the Peruvian Amazon, close to the city of Iquitos, on the banks of the Nanay River, a river now contaminated by the detritus of mining and the activities of the petroleum industry. The Munduruku chief Manoel is confined in his own ancestral territory, on the western edge of the Brazil's state of Pará, by the rapid expansion of soya monoculture. Kátia, chief of the Akrãntikatêgê people, is from the Mãe Maria Indigenous Territory, which is located near the city of Marabá, also in Pará in the Brazilian Amazon – a land scarred by mining, highways, and power lines. Each is the protagonist of other possible stories for a world anesthetized by a single narrative.

For all these reasons, *Stepping Softly on the Earth* is not simply a warning against the history of war on the Amazonian peoples, which has just entered its most frightening chapter. The film intends to be much more than this: it wants to speak of alternative horizons, where the protagonists are people who still carry an ancestral knowledge capable of showing us ways of being in the world that are completely distinct from those we have chosen as a western capitalist society, and which are leading us to ruin.

XVIII. COP26: Cognitive Dissonance[187]

This piece of reporting from COP26 in 2021 looks to those excluded from the official discussions for hope for the future.

On 3 November 2021, as I left London for Glasgow, I could see billboards and posters in Tube stations and on the streets peddling the much-hyped COP26 as the 'last best chance to save the planet.' Yet when a slogan like that stares you in the face, you are left asking: who we are saving the planet from, and for whom?

Far from the spotlights of the conference's Green and Blue Zones, the reality on the streets around the conference center is completely different and echoes the cries for more effective climate solutions.

Business in the Blue Zone

The Blue Zone is reserved for those officially registered with the UN event, who are charged with coordinating the global response to the threat of climate change. The Green Zone is where workshops, presentations, and art exhibitions take place, and is aimed at the general public.

On Friday 5 November the protest organized by the 'Fridays for Future' international climate movement and led by Indigenous peoples from a variety of countries brought together around 25,000 students and climate activists. It laid the ground for the following day's march, which raised the flag for 'climate justice' and brought more than 100,000 people together to brave the bitter winds and the damp – the *dreich*, as the Scottish say – a day without light, heat, or color. Yet the strains of the music, the chants, and the slogans not only invaded the silence of the restricted areas, to which few have access, but attracted a rich and varied throng of people into the streets, despite the *dreich*.

The two marches were able to expose the clear cognitive dissonance of COP26, which embodies the fundamental illusion that the planet's environmental problems will be solved by economic power in the form of financial investments, and technological and scientific resources, without seriously considering or analyzing the role of nature and of the peoples who inhabit the most threatened biomes. Inside the restricted rooms and controlled environments, the planet's salvation is being touted in the shape of the latest, newest products to be sold by the very same companies that are destroying it. 'Aggregating value,' 'business innovation,' and the newest 'green technology' seem to be the mantras of those who, in truth, know perfectly well that the world view underlying all these so-called new ideas is precisely what is leading us towards climate catastrophe.

No room at the table

How dare we speak of the 'planet's salvation' when we exclude from the negotiating table representatives from the countries most affected by the climate crisis, be it Brazil or elsewhere in Latin America, the Caribbean, or Africa, as well as those members of societies everywhere who are historically silenced or ignored? This deliberate exclusion avoids monitoring and pressure by the absent delegations, leaving decision-making power in the hands of governments, behind closed doors. The exclusion is implemented in the form of health restrictions, financial restrictions, or simply lack of access to the 'official zones' where discussions take place.

The question is: is COP26 just a stopgap to enable decision-makers to cool the passions of a more concerned global community who demand action – at a point when the same political and economic elites cannot even cool down the consequences of irreversible global warming?

At an event organized in the *New York Times* Climate Hub, British actress Emma Watson, activists Malala Yousafzai, Greta Thunberg, Amanda Gorman, Vanessa Nakate, Tori Tsui, Dominique Palmer, Ati Gunnawi, Viviam Misslin Izquierdo, Mya-Rose Craig, and Daphne Frias got together to discuss the power of climate activism. Thunberg stated her scepticism about the platform and declared the need to create awareness so that people around the world know what really goes on at these conferences:

> *Without massive pressure from the outside, they will continue to get away with not doing anything and just continue doing 'blah-blah-blah' and not being held accountable ... We are so far from what would actually be needed, I think what would be considered success would be if people realized what a failure this COP is*[188].

The voices raised outside the restricted areas and briefly taking over the streets of Glasgow echo the voices of those who, for thousands of years, have known how to live with and understand the Earth as a living system. They go far beyond the present transformation of everything into merchandise, and see as far ahead as our extinction. These voices point to other directions, other connections, other ideas, other possible worlds that are completely different from those offered by COP26.

XIX. Another Brazil is Possible[189]

Published in 2022, this essay focuses on an image that captures both the resilience and humanity of one Indigenous people in the face of the inhumanity and disregard of the Brazilian state.

'If a picture is worth a thousand words,' said Brazilian writer Millôr Fernandes, 'then say it with an image.'[190] This is what neurosurgeon Erik Jennings Simões did in January 2021 when he photographed the young Indigenous Tawy Zo'é, 24, carrying his 67-year-old father, Wahu Zó'é, on his back, to be immunized against COVID-19.[191] Tawy comes from the Zo'é – a recently contacted group of about 300 Indigenous who live an isolated existence in northwest Pará, near the border with Suriname, in an area that, according to the National Indigenous Foundation (FUNAI), extends over 668,500 hectares. Tawy walked more than six hours up and down steep slopes to get his elderly father, who has health problems, vaccinated[192].

The image was taken a year earlier, but was only shared by Jennings in early 2022, at which point it went viral. As the neurosurgeon himself explained in an interview with BBC News Brasil, his intention was 'to send a positive message at the beginning of the year. It was also a way of trying to send a message from the Zo'é people because they always ask if white people are getting vaccinated and if COVID-19 is over.' The symbolism of the photo of the two Zo'é is all the more powerful when we consider the lethal policy decisions

Photo: Erik Jennings

of the federal government of Jair Bolsonaro for dealing with the pandemic. In the moment captured by the camera, hope is mixed with indignation, perseverance with death.

The first impressions, fuelled by hope, were of admiration for the young man's affection and willpower. The desire to protect his father and his people makes Tawy a symbolic counterpoint to Bolsonaro, who did everything possible to discourage vaccination of the population and public health protection measures for Indigenous people. The symbolism is even more powerful considering that the word Zo'é, in the local language[193], means 'we, people like us,' as distinct from the Kirahis, who are 'outsiders, non-Indigenous.' As Tawy carries his father on his back, he carries our tiny, almost vanishing hope for a national future free from hate and bile.

Observed from an anthropological standpoint, the image makes a lie of the misconception that in Indigenous societies, the elderly are abandoned to their fate. What we see is that the elders are carried (literally) by the younger ones. It is an eloquent image of love between father and son, of moving mountains to ensure the best possible chance of preventing the older man's infection from the virus. The photo conveys the magnanimity of the gesture, which shows the significance of elders to these Indigenous peoples, to a society that frequently forgets the importance of its own.

In a still more important message, Jennings' photo testifies the scarcity of federal government resources dedicated to Indigenous people in Brazil. If Tawy was unable to carry his father, who is to say if or how the vaccine would reach him? Yet the Zo'é people, in a joint decision with the medical team who had been in the area for more than a decade, invented pandemic prevention strategies to take account of their culture, ways of life, and their protected territory.

It is well known that the government of Bolsonaro faces accusations of genocide against Brazil's native peoples. In fact, there are already at least six cases before the International Criminal Court (ICC) in the Hague, filed by groups of lawyers, Indigenous people, activists, and politicians[194]. These cases refer to the policies put into practice by the federal government, such as denying drinking water, turning a blind eye to the invasion of land grabbers and loggers, and distributing ineffective medicines to fight the pandemic. The sight of Tawy carrying his father conveys a bittersweet message. The obvious affection between father and son exposes the state's inadequacies that force the Indigenous to be advocates for themselves and their loved ones in order to enforce their rights. Despite the state, in the person of Bolsonaro, trying its best to prevent the population from getting vaccinated, Tawy and his father were immunized, together with 67.86 per cent of the Brazilian population.

The Zo'é survive by virtue of their resilience, despite the government. The dynamics of these people, who have faced many other health crises over the years, is based on their ability to adapt. It was no different during the pandemic. They had their own, unique strategy, not only for survival but also for well-being. They could not be prevented from enjoying their lives in their own territories.

XX. Epilogue: The Amazon Is Still At War

On 30 October 2022, Luis Ignacio Lula da Silva (Lula) won the presidential election for the Worker's Party (PT) by a breathtakingly narrow margin from right-wing populist incumbent Jair Bolsonaro. Despite the outgoing president's efforts, mirroring his mentor Donald Trump, to deny the result and stage a preventive coup d'etat, Lula took office on 1 January 2023. This was a night that re-awakened the possibility of dreaming. It interrupted the reign of a policy of death against humans and non-humans alike. A policy which, for four years, the Bolsonaro government had rolled out across Brazil, undermining all serious debate and leaving behind it the depredation and destruction of forests, bodies, and territories. Bolsonaro's genocidal experiment successfully employed the COVID-19 virus as a biological weapon, creating a death toll of more than 700,000 in Brazil, massively concentrated among the country's poorest populations.[195] It is good that this worst nightmare has ended. But even as we celebrate, we must also evaluate the advances, failures, and impasses of the new Lula government. Above all, in terms of what interests us here, we must look at how it has affected the largest tropical forest on the planet, the Amazon.

Take I – Lula: friend or half-friend?

In February 2020, I found myself travelling with Davi Kopenawa, his son Dario, Survival International Director Fiona Watson, and Belgian filmmaker Pieter Van Eecke (who was producing the film *Holding up the Sky* (2023) about Davi's struggle against miners on Yanomami land). During our journey through the London streets, Pieter turned to Davi and asked, 'Is Lula a friend of indigenous peoples?' Davi pondered the question and, after a long pause, replied 'He is half a friend.' To this day, I don't know if Pieter understood the depth of Davi's response. The Yanomami warrior's words still echo in my mind. Today, the question posed by Pieter and Davi's response remains: Is Lula, the Brazilian president, a friend of the Native peoples or merely a half-friend?

There is no doubt that Lula's arrival ended the ongoing extermination of the Yanomami people[196], as we discovered in the first days of the new government. At least 343 Yanomami had died due to malnutrition, malaria, and other causes, all more or less related to the presence of mining in the extensive region that stretches across the states of Roraima and Amazonas.[197] Faced with this situation, the new PT government decreed a public health emergency.[198] The Brazilian armed forces (with new commanders at the helm replacing those who supported Bolsonaro's attempted coup in January 2023), moved troops and equipment to minimize the damage caused by the previous abandonment and dismantling of Indigenous rights.

With the armed forces for the time being on its side, the new government was able to present resistance to the militias, lavishly armed through Bolsonaro's policy of uncontrolled weapons sales, who had transformed the Amazon into a lawless territory. With the return of relative security, the mainstream press could once again enter the region and report on what had happened. Meanwhile, the political leadership of the Amazon states, largely aligned with neoliberal and agribusiness interests, had to confect a public response to the new images pouring out of the region showing the desperate state of the Yanomami people.

On the other hand, in an interview with BBC Brasil, Davi's son Dario Kopenawa (currently vice-president of the Hutukara Yanomami Associação) said that the Lula government's attempts to curb illegal mining on Native lands were 'certainly a failure'[199]. If improvements in the treatment of Brazil's Native peoples had taken place, they were not enough to eliminate the starvation of Yanomami children, a further 29 of whom died in 2023[200]. Also, according to Dario, 'The mining continues. We see planes passing over our villages and we have no control. Mining continues freely on Yanomami land.'[201]

One of the most destructive measures promoted by Bolsonaro was the dismantling of IBAMA, the main federal body for the defence of the environment, which only has 782 agents operating throughout Brazilian territory[202]. One can see that, even if IBAMA was more or less saved from destruction by Lula, it is nowhere near strong enough to carry out its mission across a continent-sized land mass such as Brazil. At the very least the agency needs to be reinforced and refurbished, yet so far no steps seem to have been taken in that direction.

As this book goes to press, the largest city of the Brazilian south, Porto Alegre, has drowned under climate-change provoked flooding[203], in a disaster that rivals the 2005 devastation of New Orleans by Hurricane Katrina[204]. Brazil has already long passed the point where its environmental disasters are more or less localized in the country's backwoods and rural regions, yet environmental protection hasn't still doesn't seem to be on the Lula government's to-do list, let alone the prevention of illegal mining on Indigenous lands.

Take II – Can we begin to dream of a better future?

Saturday, October 29th, 2022, was an important day for me. This was the night that my most recent film, *Stepping Softly on the Earth*, premiered at the 46th São Paulo Film Festival[205]. In the film, I talk about three different Amazons; three stories of war. But the main characters of these stories are not defined only in terms of those who want to invade their lands. Different paths meet, different worlds merge and form stories that need to reach us . Each figure on the screen is the protagonist of other possible stories for a world that is at present bewitched by media spin and fake news.

José Pepe Manuyama is a Kukama from the Peruvian Amazon who is dealing with the contamination of the Nanay River by mining and oil. In western Pará, Chief Manoel of the Munduruku people finds his territory under siege by the expansion of soybean monoculture and export, intensified by the agribusiness interests of multinational conglomerate Cargill. Meanwhile, Chief Katia, of the Akrãntikatêgê people of Marabá (PA), struggles to maintain her culture in a territory devastated by the Vale SA mining corporation. These three stories are interconnected by the consciousness and voice of Indigenous philosopher and intellectual Ailton Krenak, who invites us to reflect on our ways of life as human beings who currently seem to insist on being on Earth by 'eating the world' in which we live.

For all these reasons, *Stepping Softly on the Earth is* not just intended to be a warning about the capitalist history of the war against the Amazonian peoples, which is now entering into its most frightening chapter.[206] The film seeks to introduce other horizons of meaning into the dialogue – horizons seen by people who, even in the present day, carry ancient concepts capable of showing us ways of being in the world that are completely different from the ruinous paths chosen by Western society. The film does not so much denounce as announce: it aims to centre our thinking around those who have historically been pushed to the edges of thought. Indigenous peoples are the protagonists of a possible future because, as Krenak (the first indigenous person to be inducted into the Brazilian Academy of Letters) predicts, the future is ancestral: we need to learn from it to step softly on the earth.[207]

The details of the destruction wrought by the Bolsonaro administration have been fully explored in the earlier chapters of this volume. Lula won in 2022, albeit by the smallest margin of difference since the end of the military dictatorship, only because, says journalist Eliane Brum, 'he had the support of a broad front that brought together everyone – from climate activists to notorious destroyers of nature – united to defeat the new fascism responsible for the loss of more than 700,000 Brazilians during the COVID-19 pandemic, where Bolsonaro carried out a virus propagation project to achieve herd immunity.'[208]

It is amidst this scene of devastation, still overseen by a congress with a hostile majority composed of pastors (the so-called 'bible caucus'), representatives of rapacious business (the 'bullet' and 'agrobusiness' caucuses), and politicized ex-military personnel, that the Lula government (called Lula-3, because he was president for two previous presidential terms from 2003–2010), is attempting to rebuild Brazil and, in terms of the global environment, protect the world.[209]

Take III – A matter of survival for humanity and the planet

Sadly, it seems that Brazil's national government, its ministers and Lula himself (like so many leaders around the world), still fail to understand that the war in defence of the Amazon is a war for human life itself. The planet's administrators cannot seem to grasp the fact that the Earth is a living system and that in this system the Amazon plays a crucial role in regulating global climate, biodiversity, and in sustaining indigenous and traditional communities. Protecting the Amazon should thus not be a local environmental issue: it is a matter of survival for humanity and the planet as a whole. Combating illegal deforestation, predatory exploitation of natural resources, and violence against the Amazon's defenders are challenges that require continuous and coordinated action from all sectors of global society in order to guarantee the long-term protection of the living treasury of the planet.

According to a study prepared by Greenpeace Brazil[210], an area equivalent to four football fields was still being devastated every day in 2023 by criminal activity in the Amazon. The destruction caused in indigenous territories alone totalled 1,409 hectares. The most alarming case comes from the Munduruku TI (Indigenous Territory), where illegal miners are breathing down the necks of 15 communities.

The principal socio-environmental impacts of these illegal activities have been widely publicized by those media sources media that cover the socio-environmental realities of the Amazon. These impacts include violence against Indigenous and traditional peoples (murders, threats, rapes, and child sexual exploitation[211]) and the cultural and social degradation of their communities; the contamination of rivers with mercury[212] which damages the health of people and animals; deforestation and the loss of biodiversity, accentuating the global climate crisis; the incursion of drug trafficking onto Indigenous lands; and the starvation that threatens many Indigenous communities .

Another study, *Continuity Scenario*[213], released in September 2023 by researchers from the Climate Center at UFRJ (Federal University of Rio de Janeiro), estimates that at the current rate, Brazil remains on course to destroy some 200,000 km² of forested area in the Amazon by 2030. Deforestation at this rate will have many grave environmental impacts. With the increase in forest loss, the Amazon could reach what scientists call a point of no return, where the forest is degraded to a point that it no longer has the capacity to regenerate and will begin an irreversible transformation into savanna.

Deforestation increases floods, fires, drought, and the loss of biodiversity. According to data from the Greenhouse Gas Emissions and Removals Estimation System, linked to the Climate Observatory network of organizations, deforestation is responsible for more than 40 per cent of gross emissions of greenhouse gasses in Brazil – gasses that are the main drivers of climate change. This, in short, is a major reason why the world is becoming hotter. 'Last year, in fact, was the hottest in at least 174 years, since meteorological measurements began,' according to the World Meteorological Organization, an agency of the United Nations.[214]

The (largely feigned) ignorance of policy-makers regarding this problem has finally begun to cause immediate and unpredictable effects all across Brazil, such as the calamitous floods that hit Rio Grande do Sul in May 2024, for the third time in less than a year.

Alongside direct deforestation (and one of its main motors) is mining, which continues largely unregulated, unrestrained and sometimes actively promoted by the Lula government. 45,065 mining concessions are currently in operation or awaiting approval in the region[215]. Of these,

21,536 overlap protected areas and indigenous lands. Mineral extraction accounts for almost 75 per cent of the exports of the Brazilian state of Pará, where elections to local authorities are often manipulated by mining interests.[216]

Studies undertaken by researchers from the National Institute for Space Research (INPE) and the University of South Alabama state that mining on Indigenous lands in the legally defined Amazon increased by 1,217 per cent over the last 35 years, jumping from 7.45 km² dedicated to this activity in 1985 to 102.16 km² in 2020. Ninety-five per cent of illegal mining is concentrated on three indigenous lands: those belonging to the Kayapó, followed by the Munduruku and the Yanomami.[217] Several new gold mines were opened in 2023.[218]

Governments, including Lula-3, like to focus on the so-called 'job creating' agribusinesses of the Amazon. Yet these have not increased the number of jobs available to the region's peoples. Between 2012 and 2019, employment in agriculture in the nine states of the Legal Amazon declined, with 322,000 jobs being lost[219]. While the average monthly wage of the sector's employees increased by 25 per cent during this period, reaching R\$829, this is still less than half the average income in the region (R\$1,692). Informal labor remains the rule: only 19 per cent of farm labourers in the Amazon (against a regional average for all employment of 40.6 per cent) are registered in the formal job market and thus have access to labor rights and social benefits[220].

In 2021, researchers from the Federal University of Pará (UFPA) published *Cattle ranching and deforestation: an analysis of the main direct causes of deforestation in the Amazon*. This study showed that livestock ranching is currently the activity causing the greatest impact upon the Amazon[221], with a correlation coefficient of 0.7345 between cattle numbers and deforested acreage. However, once land is cleared of forest, cattle are replaced by rice and corn and both of these by soya. It is common to plant rice in new areas for approximately three years before soybean production begins.[222] Thus soya and deforestation are indirectly, but strongly, linked.[223]

The Lula 3 government, in the face of resistance by politicians who defend the interests of the large soybean farmers (in some cases themselves), has indeed attempted reduce the speed with which the Amazon Forest is being destroyed. But if we consider some of the history, such as Lula's Workers' Party (PT) championing the construction of the Belo Monte Hydroelectric Plant on the Xingu river, from 2011 onwards, it becomes harder to retain any confidence in the present PT government's actions in the Amazon.

Construction of the dam (which cost around R$19 billion) 'displaced tens of thousands of people and killed local flora and fauna'[224]. Since November 2021, 'Norte Energia, owner of the dam, [has been] trying to portray itself as a well-intentioned company in order to clean up the image of a project that has been inefficient in terms of energy generation and has caused profound environmental and social impacts'[225]. Indigenous villages that resisted relocation live without electricity, even though transmission lines march straight through their territories. Norte Energia has supplied these communities with an obsolete diesel generator, 'total nonsense, as it is a very dirty form of electrical energy and much more expensive than hydroelectric power... and only works for a few hours every day"[226]. In short, either the Lula government has not yet had enough time to deal with Indigenous demands, or the criminal abandonment of these populations continues, with the aggravating factor that Belo Monte is a monstrosity, a vanity project of the various PT governments. A significant test will be whether Lula-3 allows the Volta Grande gold mine in Pará, to go ahead. Planned by Canadian mining company Belo Sun, it will be one of the largest open-cast gold mines in Latin America. The mine itself would require clearance of 2,400 hectares of forest with larger areas cleared for roads and associated infrastructure.

Each tree that dies is a world of non-human people that is incinerated with it, wrote Eliane Brum. A tree that survives does so not by itself: forests that continue to stand do not entirely depend on our pleasure. As the pandemic has shown us, as soon as we disappear from the streets, the forest, with its visible and invisible beings, begins to take everything back. The silent battle of the forest and its trees is against us and we apparently don't learn. Worse: we still actively teach what we now know to be wrong. Young women like Greta Thunberg understand that aging, angry adults make a point of not understanding.

Let us once again join and adapt our voice to that of Eliane Brum: 'we need to recover our survival instinct.'[227] If urgency is not experienced as urgency, the images of destruction that we see on television and, increasingly, outside our own windows will soon be everywhere.

Take IV – Will the Amazon forest survive?

The storm has passed, but the weather is still cloudy. In some places, it is literally pouring rain, flooding entire cities. Lives are lost without those responsible for the situation feeling the slightest remorse or embarrassment. After four years of Jair Bolsonaro 'letting through the stampede' in the Brazilian Amazon, is it possible to say today, with any certainty, that the forest has seen worse days?

During Jair Bolsonaro's reign, the deforested area of the Brazilian portion of the largest tropical forest in the world reached 35,193 km², – larger than the area of the Brazilian states of Sergipe and Alagoas combined. This was almost an 150 per cent increase over the losses in the previous governments of Dilma Rousseff and Michel Temer.

The roar of these chainsaws still echoes in the memories of many people, especially those who reside in the forest and have suffered most from the region's climate of constant violence, marked by the murders of environmental defenders, including Bruno Pereira and Dom Phillips[228].

In December 2020, the Federal Police carried out the largest seizure of illegal timber in the country's history, confiscating 226,000 m³ of logs. The operation received no praise at the time from the Bolsonaro government, whose hugely controversial Minister of the Environment, Ricardo Salles, was elected a federal deputy for São Paulo in 2022 and has since travelled to the region to provide support to the same loggers.

In the first year of Lula's third term, although there was a decrease in deforestation rates, there was also a significant increase in the number of uncontrolled fires in the Amazon.[229] In June 2023, the highest number of fires since 2007 was recorded (during the PT president's second term in office). This phenomenon is associated with the increasingly frequent droughts and heat waves caused by climate change, combined with El Niño and deforestation driven by agribusiness.

Fire is a proximate cause of deforestation.[230] While in January 2023 the number of fires decreased compared to the same month in 2022, the opposite occurred in January 2024, with a staggering 248 per cent increase over the previous year to 1.03 million ha, of which 91 per cent were in the Amazon.[231]

After the failure of the January 2023 coup, Brazilians began to discover the true legacy of the laissez faire ecological politics, anti-environmental speeches, and dismantling of public policies and agencies by Bolsonaro. The collusion of the old regime with various criminal groups in the Amazon also became clear. The mushrooming of illegal mining sites, malnourished Indigenous children, rivers contaminated by mercury, the destruction of fish stocks, assassinations carried out by criminals (often themselves very poor) could no longer be ignored.

A survey by Datafolha[232] shows that 43 per cent of Brazilians believe that the previous government's policy was focused on encouraging deforestation, the invasion of indigenous lands, and the illegal activities of miners, fishermen and hunters. It is perhaps too early to know what they believe about their new government. There are worrying signs of a revival in support for Bolsonaro, who may run for president again in October 2026. In the words of journalist Eliane Brum, who has lived in Altamira, in the eye of the hurricane, for years: 'The vast majority of Brazilians have yet to understand that the Amazon forest is not inside Brazil, but rather Brazil is on the periphery of the forest.'

Take V – Challenges and temptations

What about today? Under the command of Lula da Silva, is there once more an Amazon to dream of? The path forward will not be easy.

Numbers have, in general, improved. Deforestation fell in the first months of Lula's government compared to the same period during the Bolsonaro administration. Efforts to monitor illegal extraction have resumed. In 2023, there was a 104 per cent increase in infraction notices issued by Ibama.[233] Hundreds of pieces of equipment used by illegal miners, such as aircraft and boats, were destroyed by government agents. Although illegal mining is continuing in the Amazon, it is clear that the Brazilian State, at least, is now interested in repressing the illegalities. This might seem little to cheer about, but given the recent past, it seems like a substantial victory.

Lula's role in the effort to significantly improve Brazil's image abroad in relation to the environment is also unquestionable. The respectable presence of Marina Silva at the helm of the Ministry of the Environment (which remains, however, largely toothless thanks to the reactionary majority in Congress), constitutes an effort to restore a globally recognized Brazilian voice in defence of the forest. Lula has always publicly demonstrated his willingness to sell to world leaders the idea that Brazil would become a reference for and protagonist of the fight against climate change and in favor of environmental preservation and sustainable development. Transfers to the Amazon Fund been resumed.[234]

Problems and pressure continue, however. The Amazon is no longer the same untouched forest it once was. There is a diversity of exogenous cultures and new legitimate economic activities in the region that generate jobs and income. There are also conflicts around the exploitation of natural resources for a world that shows little interest in changing its consumption paradigms and its relationships with living and non-living beings.

The conflicts fuelled by Bolsonaro's policies remain an obstacle for Lula. 'Bolsonarism' is not dead and in various guises has thrived in Brazil for at least 500 years in the form of a bloodthirsty authoritarian extractivism. Using its congressional majority, the post-Bolsonarist coalition has hobbled Marina Silva's Ministry of the Environment and Climate Change, and the Ministry of Indigenous Peoples, led by professor Sônia Guajajara. It has weakened and revoked environmental protection laws, facilitating deforestation and the exploitation of resources in protected areas. Despite a strong reaction from the international community and civil society organizations, Lula's opponents continue to promote deregulation as a key to the country's development.

The opposition, and even some of the government's allies, are pressing for new mineral and oil exploitation across Brazil, including in the Amazon region. Many of these projects could generate conflicts with indigenous peoples and traditional communities throughout the Amazon Basin. Lula-3 government policies have even put Ibama in conflict with Petrobras, which wants to drill in the Oiapoque basin, in Amapá, on the border with French Guiana.[235] Lula says he 'continues to dream' about oil exploration in the region, despite opposition from experts and civil society.[236]

Lula is walking a political tightrope, constantly under attack by the agribusiness lobby in congress. The President himself favors the sector, as well as the Armed Forces, even though both sectors include many opponents of his policies. He often announces economic packages, incentives, financing, and credits to 'give a little help' to these sectors. Much of this is to the detriment of measures that seek to preserve the environment, instead benefiting people with interests known to be harmful to the Amazon.

In addition, the Lula-3 government needs to reverse historical problems that stem from decades of omission by the public authorities in the Amazon region, including the cumulative effects of years of deforestation. Disputes over land in the Amazon are a historical problem that generates violent conflicts between farmers, land-grabbers, indigenous people, and traditional communities. These conflicts have been exacerbated by the undermining of agrarian reform and the continuing concentration of land in the hands of a few and have become even deadlier through the militarization of the fight against organized crime.[237] As a side effect, the military presence often intimidates traditional communities, and leads to human rights violations.[238]

An example of this can be found in the Ferrogrão, an old railway project defended by ruralists and politicians involved in soybean and corn production in the Brazilian Center-West.[239] It is one of the most ambitious logistics projects in Brazil, and has often been opposed by indigenous peoples. Ironically, the Lula-3 government's agents defend the railway, which was included in the country's New Growth Acceleration Program (PAC) in August 2023. In many ways, Ferrogrão is the old Transamazonic Highway project, come to life once again. Like its predecessor, supported by the Military Dictatorship during the 1970s, Ferrogrão promises to be a dystopic dream whose attempted implementation will leave a permanent scar on the body of the forest.[240]

A similar project can be seen in the reopening of work on the BR-319 highway (Manaus-Porto Velho), which is seen as viable and as a priority for the federal government, which has even created a working group without prior consultation with the indigenous people who will be affected.

Some of these problems stem from projects of previous PT governments, like the Belo Monte dam and the New Forestry Code of 2012, passed by the PT government of Lula's ally Dilma Rousseff.

An important piece of any possible solution to this mess are the Indigenous peoples of the Amazon. Those in Brazil who suffered under Bolsonaro, and those in Amazonian countries like Ecuador and Peru, which lack even Indigenous affairs agencies comparable to Funai, and where the current presidents favour policies almost as harmful as those of Jair Bolsonaro.[241]

The future of the Amazon therefore depends on our ability to face these challenges together in a comprehensive way. President Lula's relationship with other government officials is much better than that of his predecessor. The preservation of the Amazonian biome is the only way to guarantee a prosperous future for the forest and the people who live in it. The solution is thus obvious: it is essential that we create a broad and transparent dialogue between the government, civil society, and traditional peoples in order to find lasting solutions to the conflicts which have set the Amazon ablaze.

Take VI – Indigenous peoples cannot be solely responsible[242]

The COP26 summit in Egypt in 2022 featured repeated forms of climate denialism – promises made without the capacity to deliver, band-aid solutions, and techno-salvationist fantasies that cannot countenance the possibility that white supremacist institutions and the various forms of imperialism are both unsustainable and in a state of terminal decline. It failed to consider the ways of those who live and coexist with nature, and have been doing so before the dawn of 'civilization', despite the historic and ongoing depredations of settler colonialism and necrocapitalism. Yet indigenous peoples cannot bear the sole responsibility for saving our planet.

In recent years, discussions centered around Indigenous perspectives on the environment have gained momentum, illuminating native peoples' distinctive approach to land conservation. As we contemplate the challenges faced by Indigenous communities – especially within the context of ongoing environmental crises – it is imperative to view the collective wisdom contained within their diverse cultures as a valuable tool for imagining democratic and ecologically-informed solutions for a just future for humans and all living organisms.

As we grapple with complex environmental issues, integrating Indigenous perspectives into mainstream discourse is not only a step towards justice, but a practical path forward. It's time to move beyond mere acknowledgment and actively involve Indigenous communities in shaping environmental policy and strategy. By doing so, we can collectively strive for a future where environmental protection is not a choice, but a shared responsibility borne by all.

Contrary to common misconceptions, the Indigenous peoples of the Americas do not see themselves as 'saviours' of the environment. Instead, their message is clear – without collective efforts and collaborative alliances, humanity will certainly fail to address the environmental challenges we face as a species.

While many Indigenous communities are in the vanguard of various environmental movements, it's critical to acknowledge the social wisdom informing their repeated calls for unity and solidarity across lines of national, ethnic, cultural, linguistic, and human identity. Who is willing to join forces with these peoples to think, feel, and imagine our way toward something better?

One significant aspect of Indigenous thinking highlights the importance of memory and ancestral knowledge. 'The future is ancestral. It is everything that has already existed. It's not something that is simply out there some place. It is what is right here,' says Indigenous leader, philosopher, writer, and new member of the Academia Brasileira de Letras (Brazilian Academy of Letters), Ailton Krenak.

Indigenous communities emphasize the need to preserve and embody their cultural traditions. These are by no means static: they constantly evolve in response to the pressures of history and lived experience. For the Indigenous peoples, the future is deeply rooted in the past. This worldview creates unique, complex, and highly imaginative temporal landscapes that unsettle linear notions of progress.

It is time we asked why we don't we consider the environment itself as part of our infrastructure? Indigenous voices question the prevailing paradigm that views natural resources merely as commodities to exploit. This perspective challenges us to rethink the very foundations of our societal structures and advocate for the protection of natural infrastructure.

Indigenous thinking provides an alternative framework for environmental protection that helps us to see beyond the limited purview of capitalism. It asks us to recognize the interconnectedness of all living beings and embrace a collective responsibility to safeguard our planet.

In a recent visit to Ecuador, I was able to better understand three notable cosmologies rooted in the country's diverse Indigenous philosophies: *Sumak Kawsay*, *Kawsak Sacha*, and *Tarimiat Pujustin*, all of which humbly strive to illuminate complex, dynamic relations of interdependency between humanity and nature (often tragically obscured by Western science, as well as the forms of cultural anthropocentrism that shape its pretentions of objectivity).

- *Sumak Kawsay*, popularly translated as 'Good Living', emerges from the Andes, proclaiming a life in balance and harmony between people and nature.
- In the vast Amazon rainforest, *Kawsak Sacha*, or 'Living Jungle', celebrates the spiritual connection between nature, humans, and the cosmos.
- Lastly, from the Shuar world, *Tarimiat Pujustin*, reflects a good life in their own territory and the care of nature, suggesting an intimate glimpse into Indigenous relations with the land and the more-than-human world.

These concepts converge in a call to recognize the interdependence of all forms of life and to adopt a collective approach to safeguarding our planet.

In our pursuit of solutions, we must acknowledge that Indigenous communities cannot bear this responsibility alone. Their call for collective action and mutual understanding is an invitation to forge alliances and work towards a future shared in common. Let's move beyond the rhetoric of environmentalism and actively engage with the knowledge-systems that have sustained diverse ecosystems for millennia.

Take VII – The fate of Troy

While I was writing this epilogue, I have been reading the new edition of Homer's *Iliad,* beautifully translated by Emily Wilson. I wondered whether any indigenous leader in the Amazon had ever read or heard of Homer and the fate of Troy.

Reflecting on the relevance of Homer (and specifically the *Iliad*) to modern readers and times, I wondered if Manaus or Belem could be the Amazon's Troy. Wilson wrote in her introduction:

> *Over the past century, in the wake of two world wars and amid shifting attitudes towards masculinity and heroism, The Iliad has received renewed attention… Twenty-first century readers and listeners may see the inequalities in the poem's social worlds with particular clarity. The emotional intensity of the poem's representations of conflict feels in tune with the anguish and partisanship of our era.*
>
> *We are now in a period of crisis not for a specific nation, but for humanity, inhabiting a planet that is becoming less and less habitable. A new kind of heartbreak can be felt in The Iliad's representation of a city in its last days, of triumphs and defeats and struggles and speeches that take place in a city that will soon be burned to the ground, in a landscape that will soon be flooded by all the rivers, in a world where soon, no people will live at all, and there will be no more stories and no more names… This poem will make you understand this unfathomable truth again and again, as if for the very first time[243].*

Take VI –Indigenous prophecy

Indigenous prophecies have long served as warnings about the future of humanity and the planet. In prophecy, indigenous people announce the end of the world as we know due to the destruction of the peoples of the Amazon, or warns about where we are going with the choices we've made as a society. Prophecy denounces the choices we have made that have transformed the Amazon into a global sacrifice zone, where processes destruction, and illness spread, bringing us to an end. Death, the most complete expression of the end of the possibility of life, is being brought to us by our choice to prioritize the expansion of commodity production. Our choices are plunging us into world that we don't want, but which increasingly presents itself as the only path capable of providing us a way out.

Those who best understood the creations of what we call the planet were and are the Indigenous peoples of the Amazon. That's because they are not just inhabitants of the largest rainforest on the planet, but because they were here even before the Amazon even existed as we know it. For 19,000 years, they have coexisted and cultivated the forest that, from now on, needs to be understood as a planetary biocultural heritage. There is no forest without the wisdom of the people who have always lived in it. That's why the announcement or warning of an Indigenous voice aims to save us all: it understands the possibility of the end and knows how the end is structured. But it also knows how to avoid the end.

Capitalism in the Amazon has always functioned as a war against the region's peoples. The process of generating wealth has always involved the transformation of common goods into merchandise; the transformation of Indigenous land into pasture; of a quilombola territory into an open pit mine; of peasant small holdings into large estates. Looting, defrauding, stealing, and killing have always been the verbs accompanying the arrival of capitalism in the Amazon. War is thus as the only real expression of the historical continuity of capitalism in this region.

We chose war when we chose the export of agricultural and mineral commodities over food production. We chose chaos. Across party lines in Congress, almost all the projects that were announced as 'national', have been based upon this consensus around the necessity to export primary goods. However, when we choose to make money by obliterating Indigenous people, levelling a forest, regarding nature as an obstacle, ignoring the accumulated knowledge that provided and cared for the largest and most diverse equatorial forest on the planet; when we do this, what society are we building?

Death in the Amazon rainforest is the outcome of a deliberate project. The selection of who pays, who lives, what dies is made by a few who 'own' the land, have the money, who have power. Dying in the Amazon has always been the only way to be seen by those outside the Amazon. This was the case with Eldorado dos Carajás and with the many massacres of indigenous peoples that have punctuated Amazonian history. Now, however, even death has become trivialized and no longer receives attention. We have built a society that normalizes the death of those who won't be missed, and since the Amazon never figured in the national imagination, no one misses what they can't conceive of. Death has now become as normal as capitalism in the Amazon.

If the logic of capitalist expansion and invasion in the Amazon sees the forest in the same way that a person going shopping sees the supermarket (as Ailton Krenak puts it), the Amazonian peoples offer another mindset. This uses the complementary and reciprocal relations of societies and nature to increase and produce diversity, to make food, homes, medicine, healing, paths and thus reproduce communities. The relationship with what we conventionally call 'nature' is a relationship between equals, since life is not restricted to humans. In order to restore life to the center of the debate, we cannot continue to nurture the biocultural amnesia that comes from projects that see commodity exports as the only way forward, making the Amazon a laboratory for experimenting with capitalist barbarism.

The war is continuous. Sometimes it becomes radically intense, as was the case during the Bolsonaro government. Sometimes it seems milder, as in the period we are currently in. But 'milder' is not good: it is simply the least worst. It is, at best, a corruption of our dreams that leaves us increasingly poorer, with less courage to fight for the utopias that keep us going. Degraded by the violence of the last four or six years, we accept that the boss – or the executioner – allows us to consume the forest, transformed into products, so that we, the people, become the commodity. We are at war, and the if we cannot halt it, the war will consume us.

Eliane Brum wrote:

> *It is not a war like those of the 20th century, described in history books and which haunt us to this day. Nor is it a war like the one Putin's Russia has imposed on Ukraine. It is a much longer and more decisive war: one that will extend beyond our lifetimes. When a war breaks out, one cannot choose whether to live through it or not; the war is already underway. We can only choose between fighting or waiting for the flames to engulf us while we pretending that everything is fine. Fighting alongside the peoples-nature of the Amazon in this war has nothing to do with compassion and goes far beyond ethics. This is a question of survival: we fight for life.[244] (Brum, 2024, pp. 406, my translation)*

This war is against everyone and all of us! Those who can show us ways out of this war are those who have resisted it for many centuries. Recognizing this fact and making it recognized by all is the central task of this book. As Clei Souza, the Amazonian poet, tells us in a poem that questions the narrative of those who have denied the human character of the Amazon for centuries:

> *Amazônia*
> *floresta*
> *esses que te pensam virgem*
> *não sabem que é também humana*
> *a obra do teu húmus*
>
> *esses que te pensam virgem*
> *não sabem que o teu silêncio*
> *é a soma de tantas línguas assassinadas*
>
> *esses que pensam virgem*
> *não sabem os cemitérios em tuas solidões*
>
> *não sabem eles de ajuricaba*
> *não sabem eles dos mura*
> *não sabem que pensar-te virgem*
> *é tornar-te impura.*

~~~~~~~~~~~~~

> Amazon
> forest
> those who think you are virgin
> don't know that
> the work of your humus
> is also human
> those who think you are virgin
> don't know that your silence
> is the sum of so many murdered languages
> those who think you virgin
> don't know your lonely cemeteries and rivers
> don't know about ajuricaba
> they don't know about the mura
> they don't know that to think you virgin
> is to make you impure.

Ahwatukee, Thursday 16 May, 2024

Afterword:

We are all Amazonians

By Scott Slovic

We are all Amazonians. We may look at a map and feel, at first, that the Amazonian region of South America is geographically remote from us, and yet wherever we are and whether or not we have actually travelled to the Amazon, we are breathing Amazonian air and we are implicated in the politics of destruction that are currently devastating the cultural and ecological wholeness of the region.

It is often said that the Amazonian rainforest functions as the lungs of the Earth. In truth, the planet's oceans, which are also at great risk today, produce most of the world's oxygen. But the Amazon and other rainforests are profoundly important for their role in the hydrological cycle, purifying water while also releasing oxygen into the atmosphere. The physical health of the Amazon is essential to the viability of the planet's ecosystem, but its significance goes beyond merely its material importance.

As Marcos Colón vividly reports in the twenty essays and the photographic exhibition collected here, the Amazon is ground zero for the confrontation between natural resource extraction and human lifeways that exhibit a sense of sustainable belonging to the world. In titling this volume *The Amazon In Times of War*, he is characterizing not only the acute period of violence that is occurring today in the Amazon but the less perceptible violence – what Rob Nixon calls 'slow violence' – that fuels the short-sighted comforts experienced by so many people in distant parts of the world who benefit from the products pulled from Amazonian land, such as soybeans, beef, chocolate, vanilla, rubber, gold, diamonds, wood, food coloring, pepper, palm oil, and various other staples of our modern lives.

Colón has told the history of foreign extractivist presence in the Amazon in his 2017 documentary film *Beyond Fordlandia*, particularly revealing the expansion from early twentieth-century desire for Amazonian rubber to contemporary hunger for a much wider assortment of products, chiefly a place to grow soybeans to feed the enormous population of Asia. The 'war' that Colón describes in these pages is sacrificing the ecological and cultural integrity of the Amazon, but it is actually being fought as well in our own countries, far from the quiet forest peoples and the charismatic wildlife, multicolored rivers and richly varied vegetation of the nine South American countries that share the Amazon.

Inspired by the fierce and eloquent resilience of the native communities in the Amazon, Colón has devoted himself, as a filmmaker, journalist, and scholar, to documenting both the tragic transformation of the Amazon and the beauty of what remains in this part of the world. His 2022 film *Stepping Softly on the Earth* amplifies indigenous activist Ailton Krenak's resonant phrase 'pisar suavamente na Terra' to suggest that despite the existential threats endured by the Indigenous people of the Amazon, there persists a body of knowledge and a moral will that makes these people a model for all of us throughout the world who yearn to replace our rapacious, extractivist economies with more sustainable visions of belonging to the planet and not merely subduing it in pursuit of our short-sighted material needs.

Although Colón's collection of essays and photographs may at first seem, on one level, to be journalistic project, an extension of his *Amazonia Latitude* website, this is also a beautiful illustration of the public intellectual role of the environmental humanities. Colón is a scholar and teacher of Latin American literature and film. He has studied the cultural traditions of Brazil and its neighbours for many years and devotes much of his energy to sharing these traditions with university students. But when the world is burning, both figuratively and literally, it's necessary to step out of the library and classroom and raise your voice in any way possible to call attention to the fire. Marcos Colón has devoted himself passionately and indefatigably to this effort, his own ferocity and eloquence inspired by that of the local people of the Amazon, who've welcomed him into their communities and shared their wisdom and distress with him – and by extension with all of us who watch his films and read his reports.

The question is what are the rest of us to do with the knowledge that the Amazon, one of the Earth's great cultural and ecological treasures, is so deeply threatened, is *at war*? This book is both a representation of hope of past and current struggles and an expression of hope for 'another Brazil', another Amazon. The author has long recognized that global awareness of the Amazonian plight and cooperation on solutions to the crisis are needed to avert irretrievable loss of the incalculable human and natural gifts of this beautiful, meaningful place.

Thie book is a plea to all of us in distant parts of the world to realize that we are all Amazonians, benefitting from the physical and symbolic importance of the Amazon even if we never have the opportunity to touch its trees and place our hands in its waters and watch its people dance their dances of belonging – except in films like Colón's. We are all Amazonians, and our home is at risk. What can we do with our own lifestyle choices and our written and spoken voices to join the effort Colón has so beautifully shared with us in this book?

Endnotes

[1] Described in: Torres, M. & Branford, B. (2018) *Amazon Besieged – by Dams, Soya, Agribusiness & Land-Grabbing,* Practical Action Publishing and Latin America Bureau, Rugby.

[2] There are a total of nine Amazon countries, if you include French Guiana which however, as a department of France, is not a republic.

[3] Nixon, Rob (2013) *Slow Violence and the Environmentalism of the Poor,* Harvard University Press.

[4] Krenak, Ailton (2020) *Ideas to Postpone the End of the World,* House of Anansi Press, Canada

[5] Colón, Marcos (2018) 'Le Fascisme Environnemental Hante l'Amazonie', *Mediapart,* 28 Oct 2018. https://blogs.mediapart.fr/edition/les-invites-de-mediapart/article/281018/le-fascisme-environnemental-hante-l-amazonie

Also in Portuguese at:
'Espectro Do Fascismo Ronda a Amazônia – Opinião', *Público,* 22 October 2018. https://www.publico.pt/2018/10/22/mundo/opiniao/espectro-fascismo-ronda- amazonia-1848121

And in English at:
'Environmental Fascism Haunting the Amazon', *NiCHE,* October 2018. http://niche-canada.org/2018/10/30/environmental-fascism-haunting-the-amazon/

[6] *Note:* Quilombolas are Afro-Brazilian communities descended from slaves, maintaining traditional ties to their forested territories. Originally, the 1988 inclusion of Article 68, referencing settlements of runaway slaves or 'quilombos', was perceived to address historical phenomena like the seventeenth century quilombo of Palmares. However, post-1988 actions by various stakeholders recast the quilombo concept as a reparation for slavery. Today, groups identifying with a specific historical trajectory and presumed black ancestry related to historical oppression

can seek recognition as quilombo descendants, making them eligible for collective land deeds and social welfare programs after completing a bureaucratic process.

[7]   *Note:* 'Caboclos' in Brazil refers to individuals of mixed European and Indigenous descent. Originally, during Brazil's early colonial era, the term denoted the offspring of white Europeans, typically Portuguese, and Indigenous people. Its definition has since broadened to sometimes include people in rural areas or those with a mestizo appearance. In the Amazon, rural inhabitants are often labelled as caboclos. The term, stemming from the Tupi language, has evolved into a racial and social category, distinguishing the elite from the Amazonian peasantry. Charles Wagley's *Amazon Town* offers a detailed ethnography of 'caboclo culture,' emphasizing their skills as hunters and fishermen and highlighting the urban-rural distinction.

[8]   Viga Gaier, Rodrigo (2018). 'Bolsonaro diz que pode retirar Brasil do Acordo de Paris se eleito', *Reuters,* 3 Sept 2018. https://www.reuters. com/article/politica-eleicao-bolsonaro-acordoparis-idBRKCN1LJ1YT-OBRDN.

[9]   Cowie, Sam (2019) 'Jair Bolsonaro Praised the Genocide of Indigenous People. Now He's Emboldening Attackers of Brazil's Amazonian Communities', *The Intercept*, 16 February. https:// theintercept.com/2019/02/16/brazil-bolsonaro-indigenous-land/

[10]   Ibid.

[11]   Survival International (2019) 'What Brazil's President, Jair Bolsonaro, has said about Brazil's Indigenous Peoples'. https://www. survivalinternational.org/articles/3540-Bolsonaro

[12]   e.g. https://www.businessinsider.com/amazon-rainforest-fire-brazil-bolsonaro-baselessly-blames-ngos-2019-8?op=1

[13]   Band News (2018) 'Bolsonaro: vamos botar ponto final em todos ativismos do Brasil' *Band > Portal de Notícias.* https://www.band.uol.com. br/videos/bolsonaro-vamos-botar-ponto-final-em-todos-ativismos-do-brasil-16553044

[14]   *Note:* Pantaneiros are traditional inhabitants of the Pantanal, possessing a deep-rooted culture and way of life adapted to the rhythms and cycles of this vast tropical wetland. Their heritage is a blend of Indigenous, Portuguese, and African influences, reflecting a long history of living in harmony with one of the most biodiverse regions on the planet.

[15]   *Note:* Riparian communities are groups of people settled near or alongside riverbanks and other water bodies, with their livelihoods, cultures, and daily activities deeply intertwined with the aquatic ecosystem. They often demonstrate adaptive lifestyles that harmonize with the natural rhythms and challenges of their environment.

[16] Colón, Marcos (2019) 'Política ambiental Brasileira sob suspeição', *Público*, 3 September. www.publico.pt/2019/09/03/opiniao/opiniao/politica-ambiental-brasileira-suspeicao-1885251

[17] *Note:* Chico Mendes (1944-1988) was a Brazilian rubber tapper, environmentalist and union leader who fought to preserve the Amazon rainforest and the rights of its inhabitants. His fervent advocacy against deforestation and for local communities led to international acclaim but also precipitated his tragic assassination in 1988. He remains an emblematic figure in the global environmental conservation movement. The book *Fight for the Forest – Chico Mendes in His Own Words* was published by Latin America Bureau in 1990, and is still available.

[18] *Note:* The Amazon Fund is a REDD+ mechanism created to raise donations for non-reimbursable investments in efforts to prevent, monitor and combat deforestation, as well as to promote the preservation and sustainable use in the Brazilian Amazon, under the terms of Decree N.° 6,527, dated 1 August, 2008. Managed by the Brazilian Development Bank (BNDES), the Fund primarily receives donations from international sources. In 2019 major donors Germany and Norway suspended payments to the fund, because of the increased levels of Amazon deforestation under the Bolsonaro government.

[19] *Note: The Intercept* is an American-based news organization that operates on a non-profit model. Founded with a commitment to fearless, adversarial journalism, it emphasizes in-depth investigations, unfiltered viewpoints, and rigorous reporting. The platform prides itself on its independence, refusing to be influenced by corporate interests or political bias, and seeks to bring transparency and accountability to powerful institutions through its journalistic endeavors.

[20] https://www.nationalgeographic.com/environment/article/near-brazil-amazon-fires-residents-sick-worried-angry

[21] Lovgren, Stefan (2019) 'Nas redondezas dos incêndios da Amazônia, moradores estão doentes, preocupados e com raiva', *National Geographic*, 26 August. https://www.nationalgeographicbrasil.com/meio-ambiente/2019/08/incendio- amazonia-moradores-doentes-porto-velho-rondonia-bolsonaro-queimada

[22] Borges, André, and Gabriela Biló (2019) 'Índios do Sul do Amazonas achavam que estavam livres dos incêndios acabaram dentro da catástrofe', *Estadão,* 25 August. https://www.estadao.com.br/sustentabilidade/indios-do-sul-do- amazonas-achavam-que-estavam-livre-dos-incendios-acabaram-dentro-da-catastrofe/

[23] Reuters (2019) '"Vou resistir até minha última gota de sangue", diz líder Indígena sobre queimadas no Amazonas' *Reuters*, 24 August. https://g1.globo.com/am/amazonas/noticia/2019/08/24/vou-resistir-

ate-minha-ultima-gota-de-sangue-diz- lider-indigena-sobre-queimadas-no-amazonas.ghtml

[24]   Ibid.

[25]   *Note:* The *Amazônia Legal*, or Legal Amazon, spans over 5 million km² and encompasses the Brazilian states of Acre, Amapá, Amazonas, Maranhão, Mato Grosso, Pará, Rondônia, Roraima, and Tocantins. Established by Brazilian federal law No 1.806 of 6 January 1953, this political and geographical designation was formed to usher in specific protection and developmental strategies for the Amazon region. The state of Amazonas dominates the largest portion of the Legal Amazon, covering nearly 1.6 million km², followed closely by Pará, which spans 1.25 million km².

[26]   *Note:* Founded in 2013 by Marina Silva and registered as a political party in 2015. In 2022 it joined an electoral and parliamentary group with PSOL. Silva is now Minister of the Environment and Climate Change in Lula's government.

[27]   Reuters (2019). ibid.

[28]   *Note:* Salles finally resigned in June 2021, when accused of blocking police investigations into illegal logging.

[29]   Reuters (2019). ibid.

[30]   Colón, Marcos (2019) 'Breve panorama da violência na Amazônia em 2019', *Público*, 4 December. https://www.publico.pt/2019/12/04/mundo/opiniao/breve-panorama-violencia-amazonia-2019-1896036

[31]   Nossa, Leonencio (2015) 'Favela Amazônia', *Estadão*, 5 July. https://infograficos.estadao.com.br/public/especiais/favela-amazonia/

[32]   Araújo, Thiago (2018) 'Terra sem lei: como abandono da Tríplice Fronteira Amazônica ajuda o narcotráfico no país', *Sputnik Brasil*, 10 March. https://putniknewsbr.com.br/20181003/triplice-fronteira-amazonia-trafico-12356630.html

[33]   Paiva, Luiz Fábio S. (2015) 'Nas margens do estado-nação: as falas da violência na Tríplice Fronteira Amazônica', *Revista TOMO*, No. v. 27, Revista TOMO, December 2015, pp. 327-59. doi:10.21669/tomo.v0i10.444

[34]   Ibid.

[35]   Rapozo, Pedro, et al. (2019) 'Invisibilidades e violências nos conflitos socioambientais em terras Indígenas da microrregião do Alto Solimões, Amazonas Brasil', *Mundo Amazónico*, No. 10(2), Universidad Nacional de Colombia, July, pp. 11-37, doi:10.15446/ma.v10n2.67141

[36]   Ibid.

[37]   Ibid., 13.

38   Human Rights Watch (2019) 'Mafias do Ipê: violência e desmatamento na Amazônia', *Human Rights Watch*. https://www. facebook.com/HumanRightsWatch, https://www.hrw.org/pt/video-photos/video/2019/09/23/rainforest-mafias- how-violence-and-impunity-fuel-deforestation

39   CIMI (2018). https://cimi.org.br/wp-content/uploads/2019/09/relatorio-violencia-contra-os-povos-indigenas-brasil-2018.pdf

40   Ibid.

41   Rapozo, Pedro et al. (2019). Ibid.

42   CIMI (2018). Ibid.

43   AgroCulture (2023), *AgroCulture*. https://agrocultures.org/

44   Colón, Marcos (2020) '2019 o ano em que o Brasil mais matou Indigenas', *Público*, 21 January. https://www.publico.pt/2020/01/21/opiniao/opiniao/2019-ano-brasil-matou-indigenas-1901017

45   *Note:* it may not always be clear whether a victim is a 'leader', an 'activist' or a 'community member', which makes it difficult to tally the statistics accurately.

46   *Note:* The total of killings for 2019 was 30, according to CPT,   https://www.cptnacional.org.br/downlods?task=download. send&id=14169&catid=5&m=0

47   *Note:* The Guardians of the Forest are a group of 120 Indigenous activists who are trying to protect their land, in the southeast of Brazil's state of Maranhão, which is home to approximately 12,000 members of the Guajajara and Awá peoples.

48   https://amazoniareal.com.br/no-caso-erisvan-guajajara-policia-do-maranhao-prende-inocentes-acusa-adolescente-e-nao-mostra-provas/

49   *Note:* AM denotes Amazonas. State names in Brazil are commonly abbreviated to a two-letter code. A full list of state names, abbreviations and capital cities can be found at: https://www.whereig.com/where-is-brazil/states/

50   *Note:* MA denotes Mananhão.

51   Farias, Elaíze (2019) 'Moro recebeu pedido de proteção aos Guardiões Da Floresta antes da morte de Paulo Guajajara', *Amazônia Real*, 11 November. https://amazoniareal.com.br/moro-recebeu-pedido-de-protecao-aos-guardioes-da-floresta-antes-da-morte-de-paulo-guajajara/

52   Colón, Marcos (2020) Ibid.

53   CIMI (2019) 'Relatório da violência contra os povos indígenas no Brasil-dados de 2019'. https://www.sbmfc.org.br/wp-content/uploads/2021/09/relatorio-violencia-contra-os-povos-indigenas-brasil-2019-ci-mi-1.pdf

[54]   Ibid.

[55]   Colón, Marcos (2020) Ibid.

[56]   Mendes, Karla (2019) 'Brazil's Congress Reverses Bolsonaro, Restores Funai's Land Demarcation Powers.' *Mongabay Environmental News*, 5 June. https://news.mongabay.com/2019/06/brazils-congress-reverses-bolsonaro-restores-funais-land-demarcation-powers/

[57]   Colón, Marcos (2022) 'Two Men Missing in The Amazon "wild-West"', Latin America Bureau, 10 June. https://lab.org.uk/two-men-missing-in-the-amazon-wild-west/

[58]   https://www.ibge.gov.br/cidades-e-estados/am/tabatinga.html

[59]   Nossa, Leonencio (2015) 'Favela Amazônia', *Estadão*, 5 July. https://infograficos.estadao.com.br/public/especiais/favela-amazonia/

[60]   https://ipea.gov.br/portal/publicacao-item?id=ddbbc562-1a48-4b96-a8d6-4ea639e64f57

[61]   *Note:* It is characteristic of the complexity of such areas that the same external threats sometimes unite and, as here, sometimes divide Indigenous and riverine communities.

[62]   https://www.theguardian.com/world/2023/may/19/bruno-pereira-dom-phillips-murders-brazil-marcelo-xavier

[63]   Colón, Marcos (2020) 'Ricardo Salles "Passando a boiada"': Ministro do Meio Ambiente Brasileiro muda leis ambientais na Pandemia', *Agência Envolverde*, 25 May. https://envolverde.com.br/ricardo-salles-passando-a-boiada-ministro-do-meio-ambiente-brasileiro-muda-leis-ambientais-na-pandemia/

[64]   *Note*: A GLO, or Garantia da Lei e da Ordem, is a measure allowing the President of Brazil to order the Armed Forces to intervene in a particular situation to ensure 'public security'.

[65]   *Note:* Many federal agencies underwent significant changes of role and leadership under the Bolsonaro administration: The Ministry of Education (MEC) experienced degradation with constant ministerial changes and budget cuts, including Ricardo Vélez Rodríguez, Abraham Weintraub, Carlos Decotelli, and Milton Ribeiro. The Ministry of Culture was downgraded to the Special Secretariat of Culture, with secretaries such as Roberto Alvim, Regina Duarte, and Mário Frias. The Ministry of Health saw frequent changes in ministers, including Luiz Henrique Mandetta, Nelson Teich, Eduardo Pazuello, and Marcelo Queiroga, with controversial policies regarding the Covid-19 pandemic. The Ministry of Citizenship also saw frequent changes, with Osmar Terra, Onyx Lorenzoni, and João Roma. The Ministry of the Environment (MMA) was weakened under Ricardo Salles and Joaquim Leite, resulting in increased deforestation. Agencies like IBAMA and ICMBio faced budget cuts and reduced enforcement, with Eduardo Bim at IBAMA. FUNAI

experienced budget cuts and political interference under Marcelo Xavier. INPE saw the dismissal of Ricardo Galvão after publishing deforestation data. CNPq faced severe budget cuts under João Luiz Filgueiras de Azevedo. These changes reflected a laissez-faire leadership in conflict with field agents, compromising institutional effectiveness.

66   Palheta, Nélio (2020) 'Forças armadas entram no combate aos crimes ambientais', *REDEPARÁ*, 11 May. https://redepara.com.br/Noticia/212476/forcas-armadas-entram-no-combate-aos-crimes-ambientais

67   Ibid.

68   Ibid.

69   Ibid.

70   G1 (2020) 'Agente do Ibama é ferido no rosto com uma garrafa durante ação em Uruará, no PA.', *G1*, 6 May. https://g1.globo.com/pa/para/noticia/2020/05/06/agente-do-ibama-e-ferido-no-rosto-com-uma-garrafa-durante-acao-em-uruara-no-pa.ghtml

71   *Note*: GLO: See Endnote 2, above

72   Bragança, Daniele, and Duda Menegassi (2020) 'Nanico e militarizado, reestruturação do ICMBio entra em vigor', *(O)Eco*, 12 May. https://oeco.org.br/reportagens/nanico-e-militarizado-reestruturacao-do-icmbio-entra-em-vigor/

73   Rodrigues, Douglas (2020) 'Operação das forças armadas na Amazônia custará R$ 60 milhões', *Poder360*, 11 May. https://www.poder360.com.br/governo/operacao-das-forcas-armadas-na-amazonia-custara-r-60-milhoes/

74   *Note:* The Amazon Fund is a REDD+ mechanism created to raise donations for non-reimbursable investments in efforts to prevent, monitor and combat deforestation, as well as to promote the preservation and sustainable use in the Brazilian Amazon, under the terms of Decree N.º 6,527, dated August 1, 2008. In 2019 both Germany and Norway suspended payments to the fund, because of the increased levels of Amazon deforestation under the Bolsonaro government.

75   Da Redação (2019) 'Fundo Amazônia: O anúncio do fim e o que veio depois', *Amazônia Latitude*, 24 July. https://amazonialatitude.com/2019/07/24/fundo-amazonia-o-anuncio-do-fim-e-o-que-veio-depois/

76   Maisonnave, Fabiano (2020) 'Em reforma administrativa, ICMBio centraliza e militariza gestão das Unidades de Conservação', *Folha de S.Paulo*, 13 May. https://www1.folha.uol.com.br/ambiente/2020/05/em-reforma-administrativa-icmbio-centraliza-e-militariza-gestao-das-unidades-de-conservacao.shtml

[77]   Da Redação (2019,2) 'Fogo na Amazônia: Nova crise coloca política ambiental Brasileira contra a parede', *Amazônia Latitude*, 4 September. https://www.amazonialatitude.com/2019/09/04/fogo-na-amazonia-nova-crise-coloca-politica-ambiental-brasileira-contra-a-parede/.

[78]   https://www.brasildefato.com.br/2020/05/05/os-crimes-cometidos-por-major-curio-torturador-recebido-por-bolsonaro-no-planalto

[79]   Colón, Marcos (2018) 'Will the Amazon Rainforest Become a Commodity?', *Latin America Bureau*' Latin America Bureau. https://lab.org.uk/will-the-amazon-rainforest-become-a-commodity/

[80]   *Note:* Encyclopedia.Com (2023) 'Henry Walter Bates' https://www.encyclopedia.com/people/literature-and-arts/english-literature-20th-cent-present-biographies/henry-walter-bates.

[81]   *Note:* Encyclopedia.Com (2023) 'Alfred Russel Wallace', https://www.encyclopedia.com/people/science-and-technology/biology-biographies/alfred-russel-wallace.

[82]   Bates, Henry Walter (2009) *The Naturalist on the River Amazon,* Cambridge University Press.

[83]   ANTT (1989) 'Convenção No 169 da Organização Internacional do Trabalho - Povos Indígenas e Tribais', Agência Nacional de Transportes Terrestres - ANTT, 27 June. https://portal.antt.gov.br/en/conven per centC3 per centA7cao-n-169-da-oit-povos-indigenas-e-tribais.

[84]   Locatelli, Piero (2016) 'Maicá (PA): O quilombo que parou um porto', *Brasil de Fato*, 20 June. https://www.brasildefato.com.br/2016/06/20/o-quilombo-que-parou-um-porto/

[85]   de Matos Vaz, Elizabete, et al. (2017) 'A pesca artesanal no Lago Maicá: Aspectos socioeconômicos e estrutura operacional', *Biota Amazônia*, No. v. 7, n. 4, October 2017, pp. 6-12. https://periodicos.unifap.br/index.php/biota/article/view/3168

[86]   *Note:* Empresa Brasileira de Portos de Santarém (Embraps) manages port facilities in the Santarém area of the Amazon. By directing port logistics and operations, Embraps influences the environmental and economic landscape of the Amazon basin. http://cnpj.info/Empresa-Brasileira-de-Portos-de- Santarem-Embraps

[87]   Colón, Marcos (2018) 'Will the Amazon Rainforest Become a Commodity?', *Latin America Bureau*, 22 December. https://lab.org.uk/will-the-amazon-rainforest-become-a-commodity/

[88]   Ibid.

[89]   *Note:* Cevital, an Algerian industrial conglomerate, has diversified operations encompassing agribusiness, retail, and more. The company's activities within the Amazon, particularly in agribusiness, carry environmental and economic implications for the region. https://www.cevital.com/

[90]   *Note:* Ceagro is an agricultural enterprise operating in several

regions, including the Amazon rainforest. Engaged in grain production, processing, and trade, Ceagro's activities in the Amazon are significant for both its ecological and socio-economic impact. https://www.ceagro.com/en/home/

91   See *Note, Embraps, above.*

92   *Note:* A multidisciplinary team from Universidade Federal do Oeste do Pará (UFOPA) authored this technical report, which offers a critical analysis of the Environmental Impact Study (EIA) for the Lago do Maicá Transshipment Station in Santarém (PA). Conducted under the guidelines of Secretaria Estadual de Meio Ambiente e Sustentabilidade do Pará (SEMAS) and initiated by Empresa Brasileira de Portos (EMBRAPS), the study integrates document assessments, 2016-2017 site visits, and pertinent legislative and environmental standards. The core aim is to pinpoint any deficiencies in the EIA without assessing the broader economic implications of the project.

93   Vieira, Sílvia (2018) 'Estudo técnico da Ufopa aponta "Falhas" no EIA do Projeto de Construção de Porto no Maicá', *G1*, 15 May. https://g1.globo.com/pa/santarem-regiao/noticia/estudo-tecni- co-da-ufopa-aponta-falhas-no-eia-do-projeto-de-construcao-de-porto-no-maica.ghtml

94   AguasAmazonicas.org (2022) 'Tapajós - Aguas Amazonicas', *Aguas Amazonicas*, 25 July. https://en.aguasamazonicas.org/basins/main-river-basins/tapajos

95   Ibid.

96   Locatelli, Piero (2016) Ibid..

97   Ibid.

98   Ibid.

99   *Note:* Dona Sebastiana's Account: in the documentary 'Portos em Santarém: Vítimas do Progresso,' she shares her concerns and challenges triggered by port developments in the Santarém region. She particularly underscores the adversities experienced by riparian and quilombola communities along the Tapajós riverbanks due to these capitalist initiatives. https://www.youtube.com/ watch?v=yYBjKya9lFA

100   Ibid.

101   Colón, Marcos (2018) 'A Floresta Amazónica vai se tornar uma "Commodity"?', *Público*, 22 December. https://www.publico.pt/2018/12/22/ciencia/opiniao/floresta-amazonica-commodity-1855688

102   Euclides da Cunha (1927) *Preâmbulo ao Inferno verde.* https://euclidesite.com.br/obras-de-euclides/prefacios/inferno-verde/

103   This is an edited version of an article that appeared as: 'The Amazon: Hunger, the Invisible Side of COVID-19,' Latin America

Bureau, April 16, 2020. Available at: https://lab.org.uk/the-amazon-hunger-the-invisible-side-of-COVID-19/

[104] Colón, Marcos (2020) 'Amazonia Hunger: An Invisible Vantage Point of the Pandemic', *Florida State University Digital Scholarship*, 17 March. https://fsudrs.github.io/amazonia/exhibits/amazonia-hunger/

[105] *Note:* Images of Tabatinga accompany the text in Chapter XII, below.

[106] *Note:* The exhibition is available online at: https://fsudrs.github.io/amazonia/ It can also be viewed on the LAB website https://lab.org.uk/the-amazon-in-times-of-war/

[107] Colón, Marcos (2019) 'Política ambiental Brasileira sob suspeição', *Público*, 3 September. www.publico.pt/2019/09/03/opiniao/opiniao/politica-ambiental-brasileira-suspeicao-1885251

[108] Ibid.

[109] Colón, Marcos (2020) 'The Amazon: Deregulation and Deforestation Fuel the Pandemic', *Latin America Bureau*, 26 May. https://lab.org.uk/the-amazon-deregulation-and-deforestation-fuel-the-pandemic/

[110] Danescu, Elena (2020) 'Autocratization Surges-Resistance Grows', Varieties of Democracy (V-DEM) Report, *V-Dem Institute*, Department of Political Science, University of Gothenburg. https://www.v-dem.net/documents/14/dr_2020_dqumD5e.pdf

[111] *Note:* The Brazilian *actio popularis* created in 1965 was the first instrument to give standing to any citizen to protect the public heritage. It empowers individuals to take legal action for the public good, somewhat like Judicial Review in the UK. Subsequently, in 1985, the Brazilian ACP Act, inspired by the successful US model of class action, aimed to broaden the collective rights protection to other goods, such as the environment; consumer's rights; public, social, and urban heritage; the protection of the economic order and popular economy; and the honor and dignity of racial, ethnic, and religious minorities.

[112] Bertolotto, Rodrigo (2020) 'Pandemia não ter começado no Brasil foi "Sorte", diz especialista do clima', *UOL*, 14 April. https://www.uol.com.br/ecoa/ultimas-noticias/2020/04/14/para-estudioso-do-clima-sorte-explica-pandemia-nao-comecar-pelo-brasil.htm?cmpid=copiaecola

[113] Bercito, Diogo (2020) 'Pandemia democratizou poder de matar, diz autor da teoria da "necropolítica"', *Folha de S.Paulo*, 30 March. https://www1.folha.uol.com.br/mundo/2020/03/pandemia-democratizou-poder-de-matar-diz-autor-da-teoria-da-necropolitica.shtml

[114] Colón, Marcos & Jennings, Erik (2023) 'When Healthcare is Part of the Village', *ReVista, Harvard Review of Latin America*, 17 April. https://revista.drclas.harvard.edu/when-healthcare-is-part-of-the-village/

[115] Colón, Marcos & Jennings, Erik (2020) 'Covid-19 mostra que

medicina concentrada em grandes hospitais tem que ser superada', *Folha de S.Paulo*, April. https://www1.folha.uol.com.br/ilustrissima/2020/04/covid-19-mostra-que-medicina-concentrada-em-grandes-hospitais-deve-ser-superada.shtml

116   Bolsonaro (2020) Weekly broadcast on Facebook, 23 February 2020. https://www.theguardian.com/world/2020/jan/24/jair-bolsonaro-racist-comment-sparks-outrage-indigenous-groups

117   Ribeiro, Sidarta (2020) 'Coronavirus e fascismo de Bolsonaro nos fazem esperar por nova era, diz Sidarta', *Folha de S.Paulo*, 29 March. https://www1.folha.uol.com.br/ilustrissima/2020/03/coronavirus-e-fascismo-de-bolsonaro-nos-fazem-esperar-por-nova-era-diz-sidarta.shtml

118   Schluchter W (2017) 'Dialectics of disenchantment: A Weberian look at Western modernity', *Max Weber Studies* 17(1): 24–47

119   Vila-Nova, Carolina (2019) 'Sexagenária, Revolução Cubana é ideal inconcluso', *Folha de S.Paulo*, 1 January. https://www1.folha.uol.com.br/mundo/2019/01/sexagenaria-revolucao-cubana-e-ideal-inconcluso.shtml.

120   Collucci, Cláudia (2018) 'Metade da população mundial não tem acesso aos cuidados básicos de saúde', *Folha de S.Paulo*, 23 October. https://www1.folha.uol.com.br/colunas/claudiacollucci/2018/10/metade-da-populacao-mundial-nao-tem-acesso-aos-cuidados-basicos-de-saude.shtml

121   Maisonnave, Fabiano, et al. (2018) 'Crise do Clima', *Folha de S.Paulo*. https://arte.folha.uol.com.br/ciencia/2018/crise-do-clima/introducao/

122   Valente, Rubens (2019) 'Teoria conspiratória da ditadura guia Bolsonaro na Amazônia', Ilustrissima, *Folha de S.Paulo*, 23 August. https://www1.folha.uol.com.br/ilustrissima/2019/08/teoria-conspiratoria-da-ditadura-guia-bolsonaro-na-amazonia.shtml

123   Fabrini, Fábio (2020) 'Procuradores citam risco de genocídio Indígena e cobram do Governo medidas de proteção nas aldeias', Cotidiano, *Folha de S.Paulo*, 2 April. https://www1.folha.uol.com.br/cotidiano/2020/04/procuradores-citam-risco-de-genocidio-indigena-e-cobram-do-governo-medidas-de-protecao-nas-aldeias.shtml.

124   See https://sumauma.com/en/copernico-kafka-estado-puniu-medicos-revolucionaram-saude-indigena-brasil-historia-urihi-yanomami/

125   Colón, Marcos (2020) 'Brazil's Yanomami People: Silence, Devastation and Fear', *Latin America Bureau*, 12 May. https://lab.org.uk/brazils-yanomami-people-silence-devastation-and-fear/.

126   de Andrade, Mário. Macunaíma (2023) *The Hero with No Character*, New Directions Publishing Corporation, 2023, p. 179.

127   Colón, Marcos (2020) 'Não havia ninguém lá – Yanomami podem ter o mesmo destino do povo de Macunaíma', *Público*, 1 May. https://www.publico.pt/2020/05/01/mundo/noticia/nao-ninguem-la-yanomami-

podem-destino-povo-macunaima-1914716.

128   Conselho Missionário Indigena (2018) 'Moxihatëtëa: A violência contra os povos Indígenas isolados na Amazônia e a omissão do Governo', Conselho Missionário Indigena, 1 July. https://cimi.org.br/2018/07/moxihatetea-a-violencia-contra-os-povos-indigenas-isolados-na-amazonia-e-a-omissao-do-governo/

129   Colón, Marcos (2020). Ibid.

130   de Andrade, Mário (2023) *Macunaíma: The Hero with No Character*, New Directions Publishing Corporation, p. 179.

131   Survival International (2024) 'Brazil: Crisis in Yanomami Territory, One Year after Operation to Remove Goldminers'. www.survivalinternational.org, 17 Jan. 2024, www.survivalinternational.org/news/13864

132   Ibid.

133   Zandonadi, Viviane (2023) 'Narco-Miners Challenge Government in Yanomami Territory' *Sumaúma*, 16 May. https://sumauma.com/en/narcogarimpo-desafia-o-governo-no-territorio-yanomami/

134   Survival International (2024). Ibid.

135   Colón, Marcos (2021) 'Above the "Marombas"', *ReVista*, 5 August. https://revista.drclas.harvard.edu/above-the-marombas/

136   Colón, Marcos (2020) 'Amazônia e o Enigma da "Pura Sorte"', *Público*, 2 July. https://www.publico.pt/2020/07/02/mundo/noticia/amazonia-enigma-pura-sorte-1922744

137   Reis, Nando (2020) 'Nando Reis - ensaio com novos Baianos', *Arquivo NR*. https://www.you-tube.com/watch?v=EZ5tD-49KEk

138   *Note:* See Chapter II: The Fire Balance Sheet

139   Geraque, Eduardo (2020) 'Um ano da escuridão em São Paulo: Queimadas podem tornar dia em noite?', InfoAmazonia, 19 August. https://infoamazonia.org/2020/08/19/um-ano-da-escuridao-em-sao-paulo-queimadas-podem-tornar-dia-em-noite/

140   *Note:* This was a phrase used by Ricardo Salles, Bolsonaro's environment minister, when speaking of dismantling Brazil's environmental legislation (*See* Chapter VI).

141   Thomaz, Danilo (2020) 'Como o desmatamento se alastra na Amazônia durante escalada de pandemia de Coronavírus', *Época*, 23 May. https://oglobo.globo.com/epoca/sociedade/como-desmatamento-se-alastra-na-amazonia-durante-escalada-de-pandemia-de-coronavirus-24441196

142   Menegassi, Duda (2020) 'Atraso do Governo em contratar brigadistas pode piorar cenário de queimadas em 2020', *O Eco*. https://oeco.org.br/noticias/atraso-do-governo-em-contratar-brigadistas-pode-piorar-cenario-de-queimadas-em-2020/.

143   *Note:* The Ministerio Público Federal (MPF) is the Public

Prosecutor's Office is the Brazilian body of independent public prosecutors at both the federal and state level. It operates independently from the three branches of government. Jair Bolsonao sought to undermine its independence.

144 Amaral, Ana Carolina (2020) 'Aras Nomeia defensor da "MP da Grilagem" para coordenar Câmara Ambiental do MPF – Ambiência', *Ambiência, Folha de S.Paulo*. https://ambiencia.blogfolha.uol.com.br/2020/06/08/aras-nomeia-defensor-da-mp-da-grilagem-para-coordenar-camara-ambiental-do-mpf/

145 https://www.estadao.com.br/blogs/blog/wp-content/uploads/sites/41/2020/02/Banho-de-sol.pdf

146 Lindner, Julia, et al. (2020) 'Mourão diz que crítica de investidores sobre desmatamento será respondida com "verdade e trabalho"', *Estadão*, 24 June. https://www.estadao.com.br/economia/mourao-diz-que-critica-de-investidores-sobre-desmatamento-sera-respondida-com-verdade-e-trabalho/

147 Abramovay, Ricardo (2018) *A Amazônia precisa de uma economia do conhecimento da natureza,* Alana, ISA & Greenpeace, p. 35.

148 Niranjan, Ajit (2020) 'Em meio à Pandemia, Amazônia enfrenta ameaça tripla', *Deutsche Welle*, 16 June. https://www.dw.com/pt-br/em-meio-%C3%A0-pandemia-amaz%C3%B4nia-enfrenta-amea%C3%A7a-tripla/a-53827092

149 Colón, Marcos (2020) 'Amazônia e o Enigma da "Pura Sorte"', *Público*, 2 July. https://www.publico.pt/2020/07/02/mundo/noticia/amazonia-enigma-pura-sorte-1922744

150 Colón, Marcos (2020) 'Amazônia, paraíso sob suspeição' *Revista Cult*, 26 August. https://revistacult.uol.com.br/home/amazonia-paraiso-sob-suspeicao/

151 Gondim, N. (1994) *A Invenção da Amazônia,* Marco Zero

152 Rangel, A. (1914) *Inferno Verde – Scenas e scenários do Amazonas,* Typografia Minerva. English edition, 2024: *Verdant Inferno,* Urbanomic Media Ltd.

153 Higa, Mario (2018) 'Review of Paraíso suspeito: A voragem amazónica, by Leopoldo M. Bernucci', *Luso- Brazilian Review*, vol. 55 no. 1, p. 126-130. Project MUSE https://muse.jhu.edu/article/698768

154 Nixon, Rob (2011) 'Slow Violence: Literary and Postcolonial Studies Have Ignored the Environmentalism That Often Only the Poor Can See', *The Chronicle Review*, 26 June 2011. https://www.chronicle.com/article/slow-violence/

155 Silva, M. C. (1996) 'Capitalismo e Amazônia', *Revista de Economia/ Ensaios*, Uberlândia, v. 9, n.1, p. 125-129, 1996. Also: de Freitas, M. & Silva, M.C. (2020) *The Future of Amazonia in Brazil: A Worldwide Tragedy,* Peter Lang Publishing, New York

[156]   De Andrade, M. (1928) *Macunaíma, o herói sem nenhum caráter,* Oficinas Gráficas de Eugênio Cupolo, São Paulo

[157]   Pizarro, A. (2009) *Amazonía: El río tiene voces. Imaginario y Modernización.* Fondo de Cultura Económica, Chile

[158]   Hatoum, Milton (2002) The Brothers, Macmillan.

[159]   Hatoum, Milton (2010). Orphans of Eldorado, Canongate Books

[160]   *Note:* Lula's stance (recently in the Amazon Summit) on the Amazon emphasizes its significance beyond mere exploitation, rejecting the notion of it as an empty space awaiting plunder. See: Borges, Stella, and Elck Oliveira (2023) "'Amazônia não pode ser tratada como grande depósito de riqueza", diz Lula', *UOL,* 8 August. https://noticias.uol. com.br/politica/ultimas-noticias/2023/08/08/lula-discurso-cupula-da-amazonia.htm

[161]   Ianni, Octávio (1979) *Imperialismo y cultura de la violencia em América Latina,* Siglo XXI, 1979, p. 31

[162]   Silva, M. (2012) *O Paiz do Amazonas,* Editora Valer, 3rd Edition

[163]   Benchimol, S. (1999) *Amazônia - formação social e cultural,* Editora Valer, São Paulo

[164]   Ferreira dos Reis, A.C. (1965) *A Amazônia e a cobiça internacional.*

[165]   Malheiro, Bruno Cezar (2019) 'Colonialismo interno e Estado de Exceção: A "Emergência" da Amazônia dos Grandes Projetos', *Caderno de Geografia,* No. 60, Pontificia Universidade Catolica de Minas Gerais, December, pp. 74-98. doi:10.5752/p.2318-2962.2020v30n60p74-98.

[166]   *Note:* The imaginative role of the Amazon, in this sense, could be compared with that of the East, brilliantly analyzed in Edward Said's 1978 book *Orientalism.*

[167]   Gondim, Neide (1994) *A invenção da Amazônia,* Marco Zero.

[168]   Thomaz, Danilo (2020) 'Como o desmatamento se alastra na Amazônia durante escalada de Pandemia de Coronavirus', Época, 23 May. https://oglobo.globo.com/epoca/sociedade/como-desmatamento-se-alastra-na-amazonia-durante-escalada-de-pandemia-de-corona-virus-24441196

[169]   Cólon, M. (2018) 'Olhar lento e meio ambiente: Conexões e significados muito além de Fordlândia', *Sustainability in Debate,* Vol. 9, No. 1, April, pp. 136-44, doi:10.18472/SustDeb. v9n1.2018.29861

[170]   Vitkin, M. (1995) 'The "fusion of horizons" on knowledge and alterity: Is inter-traditional understanding attainable through situated transcendence?', *Philosophy & Social Criticism,* 21 (1), 57-76. https://doi.org/10.1177/019145379502100104

[171]   Colón, Marcos (2020) 'Populations Indiennes: Seule une coalition nationale et internationale por la sécurtié sanitaire pourra les sauver', *L'Humanité,* 12 June. https://www.humanite.fr/monde/covid-19/

populations-indiennes-seule-une-coalition-nationale-et-internationale-pour-la

[172] ISA (2020). 'Médicos de Altamira e Região alertam para colapso do sistema de saúde', ISA - Instituto Socioambiental, 14 May. https://site-antigo.socioambiental.org/pt-br/noticias-socioambientais/medicos-de-altamira-e-regiao-alertam-para-colapso-do-sistema-de-saude

[173] *Note*: The Waimiri-Atroari people suffered a tragic event, often described as one of the largest genocides in Brazil during the last 40 years. This massacre occurred during the military dictatorship, where it's estimated that at least 2,000 Waimiri-Atroari Indigenous individuals disappeared, allegedly killed by state agents. The violence was linked to the construction of the BR-174 highway, intended to connect Manaus with Boa Vista. Military forces reportedly used brutal methods, including machine guns, electrified wire, bombs, and possibly chemical weapons, to clear the area for road construction. The project also involved a cassiterite mine and a hydroelectric dam, which further disrupted Indigenous territories and livelihoods. Despite the challenges of investigating events that occurred over 30 years ago, efforts are underway to uncover the truth and bring awareness of these human rights violations.
See: Branford, Sue (2013) 'Brazil -- Waimiri-Atroari Indigenous Massacre', *Latin America Bureau*, 11 June. https://lab.org.uk/brazil-waimiri-atroari-indigenous-massacre/

[174] Madeiro, Carlos (2020) 'Coronavírus seguiu rota de rios para se disseminar pela Amazônia', *UOL Notícias*, 8 June. https://noticias.uol.com.br/saude/ultimas-noticias/redacao/2020/06/08/com-barcos-cheios-coronavirus-seguiu-rota-de-rios-para-infestar-a-amazonia.htm

[175] Reinholz, Fabiana (2020) 'Subnotficação de Covid entre Indígenas mostra descaso do Governo Federal, diz comitê', *Brasil de Fato*, 15 May. https://www.brasildefato.com.br/2020/05/15/subnotifica- cao-de-covid-entre-indigenas-mostra-descaso-do-governo-federal-diz-comite

[176] Fellet, João (2020) 'Coronavírus: Moradores fogem de cidades na Amazônia para ter comida e segurança sanitária em comunidades ribeirinhas', *BBC News Brasil,* 29 May. https://www.bbc.com/portuguese/brasil-52844584

[177] Lifsitch, Andrezza (2020) 'Curva de contágio da Covid-19 cresce no AM e doença avança pra qase 90 per cent dos municípios do Interior', *G1*, 7 May. https://g1.globo.com/am/amazonas/noticia/2020/05/07/curva-de-contagio-da-covid-19-cresce-no-am-e-doenca-avanca-para-quase-90percent-dos-municipios-do-interior.ghtml

[178] Wadick, Almério, Wadick, Alves and Reis, Rodrigo (2020) 'Coronavírus se espalha e ameaça povos no Vale Do Javari', *Amazônia*

Latitude, 6 June.

[179]   Colón, Marcos (2020) 'Amazônia Redux: A Reevaluation of Urgent Needs', *ReVista: Harvard Review of Latin America*, Volume XIX, Number 3, June, pp. 81-83, https://revista.drclas.harvard.edu/amazonia-redux/

[180]   *Note:* From the author's interview with Tumi Manque Matís in Atalaia do Norte.

[181]   Acosta, Alberto, and Esperanza Martínez (2009) *El Buen Vivir: una vía para el desarrollo*, Editorial Abya-Yala.

[182]   Ostrom, Elinor (2015) *Governing the Commons: The Evolution of Institutions for Collective Action,* Cambridge University Press

[183]   Ellerbeck, Alexandra (2015) 'Sister Dorothy Stang Died Fighting for Brazil's Landless: 10 Years Later, Not Much Has Changed', *The Washington Post*, 12 February. https://www.washingtonpost.com/national/religion/sister-dorothy-stang-died-fighting-for-brazils-landless-10-years-later-not-much-has-changed/2015/02/12/c68a64a0-b2f7-11e4-bf39-5560f3918d4b_story.html

[184]   *Note:* The correct names of Indigenous groups and the locations of their communities are often debated.

[185]   Colón, Marcos (2021) 'Stepping Softly on the Earth ', *Latin America Bureau*, 23 September. https://lab.org.uk/stepping-softly-on-the-earth/

[186]   Colón, Marcos (2022). *Stepping Softly on the Earth*, Amazônia Latitude Films.

[187]   Colón, Marcos (2021) 'COP26: Cognitive Disconnections', *Latin America Bureau*, 9 Nov. https://lab.org.uk/cop26-cognitive-disconnections.

[188]   Bancroft, Hollyn (2021) 'Cop26: Political Leaders Passing Responsibility of Saving World to Young Activists', *The Independent*, 5 November. https://www.independent.co.uk/climate-change/news/thunberg-nakate-malala-cop26-panel-b1951561.html

[189]   Colón, Marcos (2022) 'Another Brazil Is Possible', Latin America Bureau, 13 January. https://lab.org.uk/another-brazil-is-possible/

[190]   Viva, Roda & Millôr Fernandes (1989). https://www.youtube.com/watch?v=A7tNSWjN0H8

[191]   Lemos, Vinícius (2022) 'A História por trás de imagem de Indígena carregando pai para se vacinar contra Covid-19', *BBC News Brasil*, 10 January. https://www.bbc.com/portuguese/brasil-59903433

[192]   *Note:* The Zo'é people live in the area between the Cuminapanema and Erepecuru rivers, located in the northwest of the Brazilian state of Pará, in the Amazon region. This Tupi-Guarani-speaking group first came into contact with outsiders in the mid-1980s. The area was officially recognized as the Zo'é Indigenous Territory in 2009 and covers 668.5 thousand hectares according to the government's Indian affairs

department, FUNAI. The territory is also known as the Cuminapanema Ethno-Environmental Front, created in 2011 (Ordinance No 1816/ PRES/, 30 December 2011). The word 'Zo'é' means 'We'. The Zo'é language belongs to sub-branch VIII of the Tupi-Guarani linguistic family, and they use this term to distinguish themselves from the kirahis, the nonindigenous. *See*: Bindá, N. H. (2001) 'Representações do ambiente e territorialidade entre os Zo'é/PA', MA dissertation, Universidade de São Paulo, São Paulo

[193]   *Note:* The language is a branch (specifically sub-group VIIIb) of the Tupi-Guarani family of languages. *See:* Rodriguez & Cabral (2002) http://www.ddl.cnrs.fr/fulltext/Rose/Rose_2012_morphologies_contact_edited_version.pdf

[194]   https://www.dw.com/en/indigenous-brazilians-accuse-jair-bolsonaro-of-genocide-at-icc/a-58810568

[195]   Nicas, Jack (2021) 'Brazilian Leader's Pandemic Handling Draws Explosive Allegation: Homicide', *The New York Times*, 19 October. www.nytimes.com/2021/10/19/world/americas/bolsonaro-covid-19-brazil.html

[196]   Nicas, Jack (2021) 'Brazilian Leader's Pandemic Handling Draws Explosive Allegation: Homicide', *The New York Times*, 19 October. HYPERLINK "http://www.nytimes.com/2021/10/19/world/americas/bolsonaro-covid-19-brazil.html"www.nytimes.com/2021/10/19/world/americas/bolsonaro-covid-19-brazil.html

[197]   Zandonadi, Viviane (2023) 'We Are Not Even Able to Count the Bodies' *Sumaúma*, 20 January. https://sumauma.com/en/nao-estamos-conseguindo-contar-os-corpos/

[198]   Colón, Marcos, and Erik Jennings (2023) 'When Healthcare Is Part of the Village', *ReVista*, 17 April https://revista.drclas.harvard.edu/when-healthcare-is-part-of-the-village/.

[199]   De Souza, O. B. (2023) 'O que você precisa saber para entender a crise na Terra Indígena Yanomami', *Instituto Socioambiental – ISA*, March 11. https://www.socioambiental.org/noticias-socioambientais/o-que-voce-precisa-saber-para-entender-crise-na-terra-indigena-yanomami

[200]   Prazeres, L. (2024) 'Yanomami: por que governo Lula não cumpriu promessa de resolver crise e o que planeja fazer agora', *BBC News Brasil*, February 8. https://www.bbc.com/portuguese/articles/crg45q8y015o

[201]   De Souza, O. B. (2023), Ibid.

[202]   Ibid.

[203]   Ionova, Ana, and Tanira Lebedeff. "Images of a Brazilian City Underwater", The New York Times, 8 May 2024, www.nytimes.com/2024/05/08/world/americas/brazil-flooding-photos.html.

204   McCoy, Terrence (2024) 'Historic Floods Kill 83, Leaving Brazil and Its President Shaken, Angry', *Washington Post*, 7 May. https://www.washingtonpost.com/world/2024/05/06/brazil-flooding-leaves-dozens-dead/

205   46a Mostra (2022) ' 46ª Mostra Internacional de Cinema Em São Paulo' 20 October. https://46.mostra.org/filmes/pisar-suavemente-na-terra

206   Medeiros, Jotabê (2022) 'Brazil: Tread Gently on the Earth' *Latin America Bureau*, 22 November. https://lab.org.uk/brazil-tread-gently-on-the-earth/.

207   Krenak, Ailton (2024) *Futuro ancestral,* Taurus.

208   Brum, Eliane (2024) *Banzeiro òkòtó: Uma viagem à Amazônia Centro do Mundo*, Primera edición, Penguin Random House, pp. 396-397 (my translation)

209   Douglas, Bruce (2015) 'Brazil's "Bullets, Beef and Bible" Caucus Wants to Imprison 16-Year-Olds', *The Guardian*, 17 April. https://www.theguardian.com/world/2015/apr/17/brazil-rightwing-caucus-lower-age-criminal-responsibility

210   Greenpeace, Brasil (2024) 'Campos de futebol por dia: Garimpo avança em terras Indígenas', *Greenpeace Brasil*, 11 March. https://www.greenpeace.org/brasil/blog/4-campos-de-futebol-por-dia-garimpo-avanca-em-terras-indigenas/.

211   Farias, Elaíze (2023) 'Facções ampliaram atuação em garimpo e em crimes ambientais na Amazônia', *Amazônia Real*, 30 November. https://amazoniareal.com.br/faccoes-na-amazonia/
Also: de Souza Pimenta, Marilia Carolina Barbosa (2024) 'Narcotráfico e garimpo ilegal na Amazônia: até onde (e como) vai o Estado?', Nexo Políticas Públicas, 7 March. https://pp.nexojornal.com.br/ponto-de-vista/2024/03/07/narcotrafico-e-garimpo-ilegal-na-amazonia-ate-onde-e-como-vai-o-estado
Also: Zandonadi, Viviane (2023) '"Narcogarimpo" desafia o governo no território Yanomami', *Sumaúma*, 16 May. https://sumauma.com/narcogarimpo-desafia-o-governo-no-territorio-yanomami/

212   Roman, Clara (2023) 'Garimpo a continua a assolar Xingu, e estrago deve perdurar por anos', *Instituto Socioambiental*, 10 November. https://www.socioambiental.org/noticias-socioambientais/garimpo-continua-assolar-xingu-e-estrago-deve-perdurar-por-anos

213   La Rovere, Emilio Lèbre (2023) 'Publicações E Diálogos', *Clima E Desenvolvimento*, 23 February. https://clima2030.org/publicacoes/

214   Folha de, São Paulo (2024) 'Entenda por que é consenso científico que ação humana causa mudanças climáticas', Projeto Comprova, 29

January https://projetocomprova.com.br/publica%C3%A7%C3%B5es/entenda-por-que-e-consenso-cientifico-que-acao-humana-causa-mudancas-climaticas/

215 Calderon, Pilar, et al. (2021) 'Informe de Evaluación de Amazonía 2021', *La Amazonía Que Queremos*, 2021. https://www.laamazoniaquequeremos.org/spa_publication/informe-de-evaluacion-de-amazonia-2021/0

216 Enríquez, Maria Amélia (2023) 'Qual a participação da mineração na geração de empregos no Pará?', *brasilmineral.com.br*, 28 August. https://www.brasilmineral.com.br/noticias/qual-a-participacao-da-mineracao-na-geracao-de-empregos-no-para

217 Mataveli, Guilherme, et al. (2022) 'Mining Is a Growing Threat within Indigenous Lands of the Brazilian Amazon', *Remote Sensing*, vol. 14, no. 16, 21 August, p. 4092. https://doi.org/10.3390/rs14164092

218 Rodrigues , Larissa , and Juliana Siqueira-Gay (2023) 'Abrindo O Livro Caixa Do Garimpo', June. https://escolhas.org/wp-content/uploads/2023/06/Sumario-Abrindo-o-livro-caixa-do-garimpo.pdf.

219 Trisotto, Fernanda (2021) 'Amazônia Legal: Crescimento da área para agropecuária não gera aumento do emprego, diz estudo', *O Globo*, 13 August. https://oglobo.globo.com/brasil/amazonia-legal-crescimento-da-area-para-agropecuaria-nao-gera-aumento-do-emprego-diz-estudo-25154323

220 Longuinho, Daniella (2023) 'Garimpos na Amazônia têm alto investimento e trabalho informal', *Agência Brasil*, 25 June. https://agenciabrasil.ebc.com.br/radioagencia-nacional/economia/audio/2023-06/garimpos-na-amazonia-tem-alto-investimento-e-trabalho-informal.

221 Rivero, Sérgio, et al. (2009) 'Pecuária e desmatamento: uma análise das principais causas diretas do desmatamento na Amazônia', *Nova Economia*, vol. 19, no. 1, pp. 41–66. https://doi.org/10.1590/s0103-63512009000100003

222 Embrapa (2023). 'Dados Econômicos - Portal Embrapa', https://www.embrapa.br/soja/cultivos/soja1/dados-economicos
Also: AgenciaBrasil (2024) 'Nova estimativa para safra de grãos na safra 2023/24 é de 295,6 milhões de toneladas', Agência Gov, February, https://agenciagov.ebc.com.br/noticias/202403/nova-estimativa-para-safra-de-graos-na-safra-2023-24-e-de-295-6-milhoes-de-toneladas.
Also: Garcia, Rafael (2021) 'Soja contribuiu para 10% do desmatamento na América do Sul em 20 anos, mostra estudo', *O Globo*, 11 June. https://oglobo.globo.com/um-so-planeta/soja-contribuiu-para-10-do-desmatamento-na-america-do-sul-em-20-anos-mostra-estudo-25054890

223 Lavor, Ana Gabriela Freire & Jullie Pereira, Thays (2024) 'Área

desmatada no Maranhão aumenta 85% em 4 anos', *InfoAmazonia*, 12 March. https://infoamazonia.org/2024/03/12/area-desmatada-no-maranhao-aumenta-85-em-4-anos-e-pressiona-terras-indigenas/

Bourscheit, Aldem (2021) 'Crescimento explosivo da boiada na Amazônia desafia corte nas emissões de metano pelo Brasil', InfoAmazonia, 9 November. https://infoamazonia.org/2021/11/09/crescimento-explosivo-pecuaria-amazonia-corte-metano-cop26/

Pereira, Jullie (2024) 'Mato Grossoe e Pará têm 25% das cabeças de gado e são maiores emissores de metano do país', InfoAmazonia, 11 January. https://infoamazonia.org/2024/01/11/mato-grosso-e-para-tem-25-das-cabecas-de-gado-e-sao-maiores-emissores-de-metano-do-pais/

Torres, M. & Branford, S. (2018) *Amazon Besieged – by Dams, Soya, Agribusiness and Land-grabbing,* Latin America Bureau & Practical Action Publishing, Rugby.

[224]   Flávio Pinto, Lúcio, et al. (2021) '*Amazonia Latitude Review* - Volume 2: Edição Especial Belo Monte 10 Anos', *Amazonia Latitude Review*, vol. 2, no. 2. https://diginole.lib.fsu.edu/islandora/object/fsu%3A781645

[225]   Ibid.

[226]   Watts, Jonathan (2014) 'Belo Monte, Brazil: The Tribes Living in the Shadow of a Megadam', *The Guardian*, 16 December. https://www.theguardian.com/environment/2014/dec/16/belo-monte-brazil-tribes-living-in-shadow-megadam

[227]   Brum, Eliane (2024), ibid, p.406 (my translation)

[228]   *See* Chapter V, above.

[229]   Agência Brasil (2024) 'Área queimada no Brasil Cresce 248% em relaçao a Janeiro de 2023', *Agência Brasil*, 27 February. https://agenciabrasil.ebc.com.br/geral/noticia/2024-02/area-queimada-no-brasil-cresce-248-em-relacao-janeiro-de-2023.

[230]   de Oliveira, G. et al. (2023) 'Increasing wildfires threaten progress on halting deforestation in Brazilian Amazonia', *Nat Ecol*, Evol 7, 1945–1946. https://doi.org/10.1038/s41559-023-02233-3

[231]   Portela, Maria Eduarda (2024) 'Amazônia tem alta de 286% nos focos de queimadas em Fevereiro', *Metrópoles*, 27 February. https://www.metropoles.com/brasil/amazonia-tem-alta-de-286-nos-focos-de-queimadas-em-fevereiro

[232]   Tavares, Joelmir (2022) 'Datafolha: 4 em 10 Brasileiros veem Incentivo de Bolsonaro a ilegalidade na Amazônia', *Folha de S.Paulo*, 25 June. https://www1.folha.uol.com.br/poder/2022/06/datafolha-4-em-10-brasileiros-veem-incentivo-de-bolsonaro-a-ilegalidade-na-amazonia.shtml

[233]   Silva, Cedê (2023) 'Amazon Deforestation down by More than One-Fifth in 12 Months', *The Brazilian Report*, 9 November. https://brazilian.report/liveblog/politics-insider/2023/11/09/amazon-deforestation-

down-by-more-than-one-fifth-in-12-month-period/

234   Marshall, Euan (2023) 'The Return of the Amazon Fund and Lula's Race to Cut Deforestation', *The Brazilian Report*, 19 January. https://brazilian.report/environment/2023/01/19/amazon-fund-cut-deforestation/

235   Saboya, Erica (2023) 'Petrobras' Plans at the Mouth of the Amazon River: What's Happening?' *Sumaúma*, 19 June. https://sumauma.com/en/exploracao-da-petrobras-na-foz-do-amazonas-o-que-e-e-quais-os-riscos/

236   Zandonadi, Viviane (2023) 'When the Tide Turns…the Slick Will Come', *Sumaúma*, 3 February. https://sumauma.com/en/quando-mare-dobrar-mancha-vai-entrar-petroleo-foz-amazonas/.

237   Gortázar, Naiara Galarraga (2023) 'Organized Crime in Brazil's Amazon Threatens the Fight against Climate Change', *El País English*, 1 December. https://english.elpais.com/climate/2023-12-01/organized-crime-in-brazils-amazon-threatens-the-fight-against-climate-change.html.

238   Dalby, Chris (2022) 'Bolsonaro vs. Lula - Dueling Visions of Crime, Security, and the Amazon in Brazil', *InSight Crime*, 29 September. https://insightcrime.org/news/bolsonaro-vs-lula-dueling-visions-crime-security-amazon-brazil/.

239   Charbel, Pedro (2024) 'Mounting Resistance to the Ferrogrão Railway in the Brazilian Amazon', *Amazon Watch*, 18 April. https://amazonwatch.org/news/2024/0418-mounting-resistance-to-the-ferrograo-railway-in-the-brazilian-amazon.

240   Ambrosio, Nicoly (2024) 'Povos Indígenas condenam a Ferrogrão', *Amazônia Real*, 12 March. https://www.amazoniareal.com.br/povos-indigenas-sentenciam-o-fim-da-ferrograo/#

241   Colón, Marcos (2024) 'Criminal Mining, Militarization and Indigenous Challenges in the Ecuadorian Crisis', *El País English*, 21 January. https://english.elpais.com/international/2024-01-21/criminal-mining-militarization-and-indigenous-challenges-in-the-ecuadorian-crisis.html.

242   *Based on* Colón, Marcos (2024) 'Indigenous peoples cannot be solely responsible' Latin America Bureau, London, January. https://lab.org.uk/indigenous-peoples-cannot-be-solely-responsible/

243   Homer *The Iliad,* translated by Emily Wilson, W.W.Norton, New York, 2023

244   Brum, Eliane (2024) *Banzeiro òkòtó: Uma viagem à Amazônia Centro do Mundo*. Primera edición, Penguin Random House, January, my translation, p.406

Index

Abramovay, Ricardo 102
Acosta, Alberto 119
Acre 120
AgroCultures international
    research network 20
Alencar, Chief Willames Machado
    23–24
Alma-Ata Declaration 84
Altamira 35, 36, 144
Alto Solimões 28–30
Amapá, state of 24, 145
Amazon Forest xxiv–xxvi, 122–
    123
    actions of IBAMA and
        ICMBio 35–37
    Amazonian reality 107
    architectural landscape of 59
    Bolsonaro's support for
        prospectors and
        loggers 39–40
    capitalist expansion and
        invasion in 151
    COVID-19 impact 108–109
    deforestation 39
    as dubious paradise 106–113
    economic and social
        development at borders
        30
    economic exploitation of 3

environmental, social, and
        cultural threats 45–48
    exploration and reparation of
        8, 102–103
    fires 9–10, 12
    global coalition for 114–116
    as green hell xxiv, 106–113
    human character of 152
    Human Rights Watch (HRW)
        report 17–19
    illegal deforestation in 19–20
    impact on Porto Velho 10
    importance of 5–6, 154–156
    INPE report 19
    'job creating' agribusinesses of
        141
    as lungs of the Earth 113, 154
    market relations 52–53
    perception of Brazilians 6
    reduction of 80
    socio-environmental conflicts
        16–17, 20
    triple border 14, 28–30
    Verde Brasil 2 operation 38
    violence in 15–19
Amazon Fund 8, 38, 144
Amazon River 58, 61, 124
Amazon Triple Border 14, 15, 28
Amazonas, state of 10, 16, 23, 28,
    29, 53, 80, 96, 97, 137
Amazonia hunger 54–55
Amazonian public health system
    97
Amazônia Real 23–24
Amazon's communities during
    COVID-19 pandemic,
    routine and life of 55–75
    Amazonian diet 74
    Belén District 68–69
    of children 73
    family photo 70
    local fishing scene 68, 71–74

lunchtime 71
marketplace 73
material life 65
means of maintaining
    livelihood 70
means of transport 63
reliance on river's resources
    66–67
sunset 65, 71
trade 64
washing of clothes 72
Ananais River 55, 71
Andrade, Mário de 90, 107
Angamos community 58
Anísio Jobim Penitentiary
    Complex (COMPAJ) 13
'another Brazil' 155
Araguaia Guerrilla War 39
Aras, Augusto 102
Association of Civil Servants
    in Environmental
    Management (ASIBAMA/
    PA) 36–37

BAPEs (Ethno-Environmental
    Protection Bases) 91–92
Barros, Edmar 32
Bates, Henry Walter 42
    The Naturalist on the River
        Amazon 42
BBC News 132
Belém xx, 35, 78, 149
Belén District 68–69
Belo Monte Hydroelectric Plant
    141–142
Bergamo, Mônica 12
Bernucci, Leopoldo 106
Beyond Fordlandia 154
Bim, Eduardo Fortunato 36
Blue Zone 128–129
Boa Vista city 82

Bolivia 3, 29, 124
Bolsa Família (the Family
    Allowance) 6–7
Bolsonarism 145
Bolsonaro, Jair , xxiii–xxiv, 2, 11,
    35, 82–83, 97, 101, 119,
    133, 137, 139, 143–144,
    146
    as Brazilian Trump 3, 6–7
    on Brazil's sovereign right and
        duty 5
    on demarcation of Indigenous
        lands 3–4, 11, 20
    environmental fascism of 4,
        6–7
    idea of progress 2
    national interests 5
    protection of Indigenous
        culture 25
    public position on Indigenous
        territories 31
    restriction on land use 4
    support for prospectors and
        loggers 39–40
Brasil de Fato 46
Brazil 59, 124
    on Amazon 120–121
    Brazilian Constitution of 1988
        4–7, 76
    combating deforestation 19
    drug trafficking 14, 29–30
    environmental policy 24–25
    Mercosur agreement and
        embargos on 9
    mineral and oil exploitation
        across 145
    National Guard 10
    necropolitics 78
    New Forest Code 19
Brazilian Academy of Letters 139,
    147
Brazilian *actio popularis* 167n114

Brazilian Forest Service (SFB) 35,
    101
Brazilian Institute of Geography
    and Statistics (IBGE) 29
Brazilian Institute of the
    Environment and
    Renewable Natural
    Resources (IBAMA) 6, 8,
    34–38, 76–77, 102, 137
Brazilian New Tribes Mission
    (MNTB) 91
Brazilian society 120
BR-179 highway 116
Brum, Eliane 139, 142

Caballococha 51, 67
Caboclos 2, 6–7, 159n7
Cacau Pirêra community 96–97
capitalism in Amazon 149–152
Cardona, Father Guilherme 46
Careiro da Várzea 98, 99
Cargill 44–45, 138
Carvalho, Juliano Baiocchi Villa-
    Verde de 102
CEAGRO company 45
Cemitério dos Índios (Indian
    Cemetery) 23
    land invasion 24
Chico Mendes Institute for
    Biodiversity Conservation
    (ICMBio) 6, 8, 34–37, 77,
    77
Chile 28
climate denialism 146
climate justice 128
Climate Observatory 140
Coalition of Indigenous People
    of the Northeast, Minas
    Gerais and Espírito Santo
    (APOINME) 26

Coalition of Indigenous Peoples
    of Brazil (APIB) 25
Cochiquinas community 60
Colombia 3, 14–15, 28–29, 31, 51,
    66, 108, 124
Colón, Marcos 154–155
colonialism 6–7, 120, 147
colonial maps 124–125
Comando Vermelho (CV) 14, 23
Comissão Pastoral da Terra
    (Pastoral Land
    Commission) (CPT) 22
Comissão Pró Yanomami (CCPY)
    86
conservation units 19, 37, 77, 78
Continuity Scenario study 140
Convention 169 of the
    International Labour
    Organization (ILO) 4
Coordination of Indigenous
    Organizations of the
    Brazilian Amazon
    (COIAB) 115–116
COP26 128–129, 130, 146
COVID-19 pandemic 50–51, 56,
    59, 63, 85, 92, 98, 115,
    115, 123, 139
    economic impact of 93
    impacts 80–81, 97
    vaccination 132–134
Craig, Mya-Rose 129
creative invention 107
culturality 84–85
Cunha, Euclides da xxiv, 48
Curió, Sebastião 39–40

Datafolha 143
deforestation 14, 39, 79, 139, 140,
    143, 144
    fire as cause of 143
    illegal 80–81

during Lula's government
    143–144
proliferation of disease and 80
demarcation of Indigenous lands
    3–4, 11, 20, 26
    Executive Orders and Proposed
        Constitutional
        Amendments (PECs)
        26
Deregulation 34
Dias, Ricardo 91
Dias, Tereza Cristina Corrêa da
    Costa 35
Diniz, Edvaldo 98
dreich 128
drug trafficking 14, 29–30

economic zoning 4–5
Ecuador 3, 28, 124, 146, 148, 148
Eecke, Pieter Van 136
El Niño 143
Eldorado dos Carajás 150
Embraps 45, 46
environmental crisis 9–10
environmental degradation xxiii,
    79
environmental fascism 4, 6
environmental policy 24–25,
    102–103
    actions of IBAMA and
        ICMBio 35–37
    agribusiness interests 35
    Decree No. 10.347 35
    environmental preservation and
        regulations 34
    Forest Code (Código Florestal)
        53
    government's use of GLO 34,
        37–39
    management of public forests
        35, 78–79

measures of deregulation 34
use of armed forces 37–39, 77
Verde Brasil 2 operation 38
Época 101
Erlan, Bruno 51
Ethno-Environmental Protection
    Bases (BAPEs) 92
European Union 8, 9

Família do Norte (FDN) 14–15,
    23, 29
Federal Prosecutor's Office (MPF)
    88, 92, 102
Federal University of Pará (UFPA)
    141
Federal University of Rio de
    Janeiro (UFRJ) 140
Fernanda, Maria 56, 59
Fernandes, Millôr 132
Fernandes, William 36
Fila Del community 63
floods 97–98, 140
Folha de São Paulo 11–12
Fome Zero (Zero Hunger) 6–7
Forest Code (Código Florestal) 53
forest management, regulations 35,
    78–79
Forestry Grant Plan, Annual
    (PAOF) 78
Frias, Daphne 129
'Fridays for Future' international
    climate movement 128

Galvão, Ricardo 101–102
garimpeiro (land-grabber) 145
genocide xxiii, 76, 92, 94, 97, 103,
    114–115, 133, 172n176
Gondim, Neide 106
Gorman, Amanda 129
green hell xxiv, 106–113
Greenpeace Brazil 139–140

Green Zone 128
Grupo Cevitai 45
G7 summit agenda 9
Guajajara, Célia Lúcia 23
Guajajara, Erisvan Soares 23
Guajajara, Firmino Prexede 24
Guajajara, Paulo Paulino 23, 24
Guajajara, Raimundo Benício 24
Guajajara, Sônia 85, 145
Guarantee (GLO) decree 34
Guardians of the Forest 24
Guianas 124
Gunnawi, Ati 129
Guyana 3

Hatoum, Milton 108
Holding up the Sky 136
Homer's *Iliad* 149
House of Indigenous Health
          (CASAI) 82
Human Rights Watch (HRW) 17
Hutukara Yanomami Association
          (HAY) 93
hydroelectric
     dam 47
     plant 91, 116, 141

Ianni, Octávio 108
IBAMA —see Brazilian Institute
     for the Environment and
     Renewable Natural Resources
illegal mining 28, 77, 81, 87, 91, 93,
          137–138, 141, 143–144
Indigenous healthcare practices
          82, 114
     community health centers
          85–86
     concept of 'culturality' 84–85
     forms of knowledge and
          interventions 84–85
     Urihi system 86

Yanomami experience 82–84
Zo'é model of healthcare
          86–88
Indigenous lands
     demarcation of 3–4, 11, 20, 26
     environmental protection
          76–78
     land-grabbing and prospecting
          operations on 36
     land-use changes 53
Indigenous Missionary Council
          (CIMI) 17–18, 91, 115
     Support Team for Isolated
          Indigenous Peoples 26
Indigenous people 155
     acquisition of new food
          products 53–54, 56–60,
          62–63
     agricultural communities 3
     Angamos community 58
     fishing communities 53
     genocide against 133, 150
     imaginary narratives and
          representations 112
     Indigenous struggle 25–26
     private property holdings and
          title deeds 76
     religious group 57
     routine and life of 60, 63
     as 'saviours' of the
          environment 146–147
     shooting of Indigenous leaders
          23–24
     silence of 119–121
     Tabatinga misery 14, 15
     Tenharin 10
     territory of 3–4, 3–4, 11
     traditional literary vision of 109
     violence against 17–18, 20,
          22–23
     vulnerability to diseases 81

Indigenous rights 26, 31–32, 76, 83, 111–112, 115, 122, 133, 136–137, 141, 160n20
Indigenous thinking 147–148
informal labor 141
Institute for Applied Economic Research (IPEA) 30
Institute for Environmental Protection (PROAM) 11
international civil society 115
International Criminal Court (ICC) 13
international economic partnerships 8
International Seminar on Political Ecology 20
invaders, land 101
Iquitos 56, 59, 66, 72, 125–126
    Water Defense Committee of 125
isolated
    Indigenous peoples 26, 82, 91, 132
    Indigenous communities 50, 53, 116
    Indigenous populations 54
Itaquaí River 31
Itaya River 56
Ituna–Itatá Indigenous Territory 77
Izquierdo, Viviam Misslin 129

Javari Valley 31–32, 116, 121
João Lira 47
Johnson, Boris 9
Juma community 99

Kawsak Sacha 148
Kopenawa, Dario 137
Kopenawa, Davi 93, 136, 137
Krenak, Ailton xxvi, 122, 125, 155

Lago do Maicá 42–43
    environmental impact study of 45
    hydrographic basin 44
    important for archaeology 43
    as natural breeding area 43
    port zone in 44
    traditional peoples and communities living in 45
land grabbing –see garimpeiro
land invasions in Amazon 18, 20
land rights 18, 20, 35, 35
Law and Order Guarantee decree (GLO) 10, 34, 37–39, 77, 81
Legal Amazon 11, 77, 141
Lima, Divanildo dos Santos 36
loggers 10, 17, 19, 24, 36 – 39, 82, 133, 143
Lula –see Silva, Luis Inácio Lula da

Macron, Emmanuel 9, 10
Macunaíma 90, 93
Mãe Maria Indigenous Territory 126
Madeira River 10, 43
Manaus 108
    Free Trade Zone 120
    Moderna Market 98
Manoel, Cacique 124, 138
Manuyama, José Pepe 125, 138
Maranhão, state of 23,24
marombas 96, 98–99
Martins, Pedro 44
Matís, Tumi Manque 118
Mato Grosso, 38, 44, 46, 47, 120
Matos, Jackson Rêgo 45–46
Matsés tribe 118
Mbembe, Achille 81
Mendes, Chico 8

Mercosur–European Union agreement 8–9
military dictatorship 5, 39, 139, 146
Mining
  Gold mining xxi, 90-94
  Miners 24, 123, 124
  Illegal 28, 77, 80, 87, 137, 138, 143
Ministry of Agriculture, Livestock and Food Supply (MAPA) 34
Ministry of Environment (and Climate Change) (MMA) 35
Ministry of Indigenous Peoples 145
Missionaries 91
Moro, Sérgio 24
Mourão, General Hamilton 38, 102
Munduruku Indigenous people 123, 124, 126, 138, 141
  TI (Indigenous Territory) of 19, 141
Mura, Raimundo 10–11
Mura people 10–11, 24

Nakate, Vanessa 129
Nanay River 124, 138
National Council of the Amazon (CAN) 35, 38, 102
National Fund for Forest Development (FNDF) 35
National Indigenous Foundation (FUNAI) xxiv, 8, 25, 32, 35, 36, 76–77, 86–88, 91, 92, 132
National Institute for Space Research (INPE) 8, 19, 39, 101, 141

New Forestry Code of 2012 19, 146
New Growth Acceleration Program (PAC) 146
New York Times 1
Nixon, Rob xxiii, 107, 154
Nobre, Carlos 79
Norte Energia 142
Northern Arc 28, 30
NOVO 11

O Eco 102
O Estado de S.Paulo newspaper 12, 14, 102
'Onça Preta' ('Black Jaguar') 23
Organization for Economic Cooperation and Development (OECD) 8
Os Pingo da Chuva ('The Raindrops') 100
Ostrom, Elinor 119
Oswaldo Cruz Foundation 115
'otherness' 109, 113

Paiva, Luiz Fábio 15
Palmares Cultural Foundation 43
Palmer, Dominique 129
Pantoja, Mário 46
Pará, state of 37, 39, 43, 45, 47, 83, 86, 121, 122, 126, 133, 139, 141, 142
paradise, idea of 106
Paris Climate Agreement 3, 19
Paris of the Tropics 97
Pastoral Land Commission (CPT) 17
Pebas community 58
Pereira, Bruno Araújo 31–32, 143
Peru 3, 3, 14, 28–29, 31, 108, 124, 146
Phillips, Dom 31–32, 143

Pizarro, Ana 107
Porto Alegre 138
Porto Velho 11
Prev-Fogo (Prevent Fire) program
    102
Primeiro Comando da Capital
    (First Capital Command)
    (PCC) 14, 29
Prospectors 10, 17, 24, 38, 82
Purus River 11
Py-Daniel, Anne Rapp 43

Quilombolas 2, 7, 17, 18, 43, 44,
    46, 48, 119–120, 151, 158

Rangel, Alberto xxiv, 106
Rapozo, Pedro 16, 18
Real-Time Deforestation
    Detection (DETER) 19
REDE (the Sustainability
    Network) 11
Reis, Nando 100
Rio Grande do Sul 140
Riverside/riverine communities 31,
    50-53, 65, 119
Rodrigues, Randolfe 12
Rondônia, 11
Roraima, state of 82, 90–93, 137
Rousseff, Dilma 143, 146
rubber boom xx

Safra Plan 102
Salles, Ricardo xxiv, 8, 11–12, 34,
    143
Santarém xx, 42–48, 123, 125
Satellite Deforestation Monitoring
    Project for the Legal
    Amazon (PRODES) 19,
    101
Seabra, Abrahim 98, 99
Sebastiana, Dona 47

Sena, Edilberto 44, 47
Serra Pelada 40
Silene, Kátia 123
Silva, Corrêa da 108
Silva, Fernando Azevedo e 38
Silva, José Roberto do Nascimento
    23
Silva, Luis Inácio Lula da, 'Lula',
    xxiii, 82–86, 88, 94, 137–
    139, 144–147
Silva, Marilene Corrêa da 107
Silva, Marina 144–5
Simões, Erik Jennings 132
slavery xxiii, 109, 112
Solimões River 50, 75
Souza, Carlos Alberto Oliveira de
    24
Souza, Clei 151
soya xx, 126, 141
soybean cultivation and export
    43–44, 123, 155
Special Secretariat for Indigenous
    Health (SESAI) 19, 116
Squatters 17, 38
Stang, Sister Dorothy 120
Stepping Softly on the Earth 126, 138,
    155
Sumak Kawsay 148
Supreme Federal Court (STF) 35
Suriname 3, 124
Survival International 94–95

Tabatinga village 28–30, 57, 59, 61,
    73, 108
    food supply 51–52
    misery 14, 15
    seasonality and migration
        52–53
Tapajós River 43, 46, 47
    –Teles Pires axis 44
Tarimiat Pujustin 148

Temer, Michel 3, 143
Tenharin, Antônio Enésio 10
Tenharin Indigenous people 10
terra nullius 108
Thunberg, Greta 129, 142
Trans-Amazonian highway xx,1223
Trump, Donald 3
Tsui, Tori 129
tuberculosis 80, 94
Tupiniquim, Paulo 25–26

uncivilized barbarians 109
Union of Indigenous Peoples
        of the Javari Valley
        (UNIVAJA) 31–32
Uraricoera River 90
Urihi Saúde Yanomami 86

Vale Mining Company 123
Venezuela 3, 30, 52, 82, 95, 124
Ventura, Luis 91
Verde Brasil 2 (Green Brazil 2)
        operation 38
violence 151
    in Amazon forest 15–20
    CIMI report 18
    with impunity 17
    against Indigenous people
            17–18, 20, 22–25
    lines of investigation 23
    slow xxiii, 154
Vizcarra, Martin 50

Waimiri-Atroari people 115–116,
        172n176
Wajãpi, Emyra 23–24
Wallace, Alfred Russel 42
Watson, Emma 129
Watson, Fiona 94–95, 136
West Pará Federal University
        (UFOPA) 45

Wilson, Emily 149
World Health Organization
        (WHO) 79
World Meteorological
        Organization 140

Xavier, Marcelo 32
Xingu River 141

Yanomami Home for Indigenous
        Health (CASAI) 82
Yanomami Indigenous Territory
        82–83, 90–93
    area 90
    ecological problems 93–95
    humanitarian crisis 94
    illegal gold mining 91, 93
    pandemic-related
            unemployment 93
    religious missions 91
Yanomami shamans 86
Yanomami people
    ancestral way of life 82
    healthcare experience 82–84
Ye'kwana Indigenous peoples
        90–91
Yousafzai, Malala 129

Zo'é, Tawy 132, 134
Zo'é, Wahu 132, 134
Zo'é people 132
    healthcare experience 86–88
    indigenous lands 87–88